Sustainable UAV Ecosystems

Sustainable UAV Ecosystems: Integrating Blockchain, Quantum, and AI for Smart Mobility and SDG Innovation provides a comprehensive examination of technologies that enable secure, intelligent, and sustainable UAV ecosystems for smart mobility and digital infrastructure. It presents theoretical foundations and practical methodologies integrating artificial intelligence, blockchain, post-quantum cryptography, and cloud-edge computing for resilient UAV communications. It not only explores predictive intelligence, adaptive routing, swarm intelligence, and decentralized control in autonomous aerial networks but also highlights applications in smart-city logistics, disaster response, environmental monitoring, and precision agriculture. Emphasis is placed on energy-efficient operations, low-carbon mobility, and sustainable UAV systems aligned with the Sustainable Development Goals (SDGs).

Key features:

- Illustrates the design and integration of secure and sustainable UAV ecosystems, including intelligent aerial networks, smart mobility infrastructures, and autonomous drone operations.
- Emphasizes the role of artificial intelligence and intelligent analytics in UAV systems by leveraging advanced techniques such as predictive intelligence, adaptive routing, and swarm intelligence for cooperative aerial operations.
- Presents practical applications of UAV technologies in smart-city logistics, disaster response, environmental monitoring, and precision agriculture to support efficient and sustainable services.
- Covers cybersecurity mechanisms and post-quantum cryptographic models for protecting UAV communications, aerial networks, and decentralized drone infrastructures.
- Showcases the integration of blockchain, distributed cloud-edge computing, and decentralized control frameworks to enable trustworthy, scalable, and resilient UAV ecosystems.

This book is suitable for senior undergraduates, graduate students, and academic researchers in computer science, artificial intelligence, aerospace engineering, robotics, communication engineering, and smart mobility systems, as well as policymakers and industry practitioners working on sustainable UAV technologies and intelligent aerial infrastructures.

Sustainable UAV Ecosystems

Integrating Blockchain, Quantum, and AI for Smart Mobility and SDG Innovation

Edited by
Hemant Kumar Saini,
Yassine Maleh, Sita Rani,
and Surbhi Bhatia Khan

CRC Press
Taylor & Francis Group
Boca Raton London New York

CRC Press is an imprint of the
Taylor & Francis Group, an **informa** business

Designed cover image: Shutterstock ID: 535282921

First edition published 2027
by CRC Press
2385 NW Executive Center Drive, Suite 320, Boca Raton FL 33431

and by CRC Press
4 Park Square, Milton Park, Abingdon, Oxon, OX14 4RN

CRC Press is an imprint of Taylor & Francis Group, LLC

ISBN: 978-1-041-22288-0 (hbk)
ISBN: 978-1-041-22289-7 (pbk)
ISBN: 978-1-003-73158-0 (ebk)

DOI: 10.1201/9781003731580

Typeset in Sabon
by SPi Technologies India Pvt Ltd (Straive)

Contents

About the Editors

Hemant Kumar Saini works as an academic at the School of Computer Science Engineering at Bennett University (The Times Group), Greater Noida, India, and Lincoln University, Malaysia. He has more than 14 years of teaching experience. His research includes digital transformation technologies, such as aerial vehicles, ad-hoc networks, artificial intelligence (AI), quantum computing, Internet of Things (IoT), blockchain, edge and serverless computing, cloud-native computing, and digital twins.

Yassine Maleh received his PhD from the Université Hassan 1er in Morocco in the field of Internet of Things security and privacy in 2013. He is a senior member of IEEE and a member of the International Association of Engineers, IAENG, and the Machine Intelligence Research Labs. Dr. Maleh has made contributions in the fields of information security and privacy, Internet of Things security, and wireless and constrained network security. His research interests include information security and privacy, the Internet of Things, network security, information systems, and IT governance.

Sita Rani works in the Department of Computer Science and Engineering at Guru Nanak Dev Engineering College, Ludhiana. She has more than 20 years of academic, administrative, and research experience. Her research interests include aeroponics, artificial intelligence, machine learning, the Internet of Things, blockchain technology, smart healthcare, and sustainable development.

Surbhi Bhatia Khan is working in the School of Science, Engineering and Environment at the University of Salford, United Kingdom. She is a project management professional certified by the PMP, USA. She is ranked among the top 2% of scientists globally for the second year in a row (2023 and 2024). She is supervising/co-supervising a number of PhD and research associates. She has published over 250 articles and books in high-indexed outlets with her students and colleagues and has also been granted patents from the USA. She is also serving as a deputy EIC in TCE letters, an associate editor in IEEE Transactions on Consumer Electronics, HCIS and EAAI journals, and a guest editor in many other reputed journals. She is working in KTP Projects Innovate UK and has submitted grants as a CO-PI inHORIZON-EIC-2025-PATHFINDEROPEN. Her research interests are information security and privacy, quantum computing, machine learning/deep learning, data science in healthcare and IoT, text analytics and detection, and sentiment analysis.

Contributors

Sarumi Usman Abidemi
Department AI, Data Analytics, and Software Engineering Departments
AI and IoT Research Center
AI and Informatics Faculty
Near East University
Mersin, Turkey

Fadi Al-Turjman
Department AI, Data Analytics, and Software Engineering Departments
AI and IoT Research Center
AI and Informatics Faculty
Near East University
Mersin, Turkey

Sinem Alturjman
Department AI, Data Analytics, and Software Engineering Departments
AI and IoT Research Center
AI and Informatics Faculty
Near East University
Mersin, Turkey

P. R. Anisha
Faculty of Engineering and Technology
Stanley College of Engineering and Technology for Women
Hyderabad, India

J. Balaraju
School of Engineering
Anurag University
Hyderabad, India

Chandradeep Bhatt
Department of Computer Science and Engineering
Graphic Era Hill University
Dehradun, India

Raja Rao Budaraju
Oracle America Inc
Dublin, CA, United States

Susheela Dahiya
Department of Computer Science and Engineering
Graphic Era Hill University
Dehradun, India

C. Dastagiraiah
School of Engineering
Anurag University
Hyderabad, India

Srinath Doss
Faculty of Engineering and Technology
Botho University
Botswana, Southern Africa

Enugurthi Harika
School of Engineering
Anurag University
Hyderabad, India

Tanusha Gorak
Faculty of Engineering and Technology
Stanley College of Engineering and Technology for Women
Hyderabad, India

Sandeep Gupta
Department of Artificial Intelligence and Data Science
Poornima Institute of Engineering and Technology
Jaipur, India

M. Shamim Kaiser
Institute of Information Technology
Jahangirnagar University
Dhaka, Bangladesh

Rishi Kant
School of Allied and Healthcare Sciences
GNA University
Punjab, India

Gurleen Kaur
Faculty of Applied Sciences
Sheridan College
Ontario, Canada

Gangaraju Krishna Keerthana
Department of CSE
Siddhartha Institute of Technology and Sciences
Hyderabad, India

Udit Mamodiya
Faculty of Engineering and Technology
Poornima University
Jaipur, India

Cliford Peter Mlimwa
Department AI, Data Analytics, and Software Engineering Departments
AI and IoT Research Center
AI and Informatics Faculty
Near East University
Mersin, Turkey

Praveen Kumar Mundlagiri
School of Engineering
Anurag University
Hyderabad, India

Sri Nagesh Osri
School of Engineering
Anurag University
Hyderabad, India

Vijaishree Dubey Pandey
School of Law
CHRIST University
Delhi, India

Manpinder Singh Panesar
Amazon Web Services
Ashburn, Virginia, United States

Jothi Paranthaman
Faculty of Engineering and Technology
Botho University
Botswana, Southern Africa

Pasula Sowjanya
Department of Cybersecurity
Sreenidhi Institute of Science and Technology
Hyderabad, India

Balusupati Veera Venkata Siva Prasad
School of Engineering
Anurag University
Hyderabad, India

Kamalakar Ramineni
School of Engineering
Anurag University
Hyderabad, India

Simar Singh Rayat
Department of Computer Science and Engineering
Graphic Era Hill University
Dehradun, India

C. Kishor Kumar Reddy
Faculty of Engineering and Technology
Stanley College of Engineering and Technology for Women
Hyderabad, India

Riya
University Institute of Liberal Arts & Humanities
Chandigarh University
Punjab, India

Juvariya Sadaf
Department of Pharmacy
State Board of Technical Education and Training (SBTET)
Andhra Pradesh, India

Rimsha Sadaf
Faculty of Engineering and Technology
Stanley College of Engineering and Technology for Women
Hyderabad, India

Hemant Kumar Saini
Department of CSE
Lincoln University College
Petaling Jaya, Malaysia

Sahil Sharma
School of Allied and Healthcare Sciences
GNA University
Punjab, India

Vikrant Sharma
Department of Computer Science and Engineering
Graphic Era Hill University
Dehradun, India

Nishi Kumar Shukla
Wimpey Laboratories
Abu Dhabi, United Arab Emirates (UAE)

Sheetal Singh
School of Law
CHRIST University
Delhi, India

Rituraj Sinha
School of Law
CHRIST University
Delhi, India

Shashank Solanki
School of Law
CHRIST University
Delhi, India

B. V. Srikanth
School of Engineering
Anurag University
Hyderabad, India

Bebesh Tripathy
Department of Computer Science & Engineering, UIE
Chandigarh University
Punjab, India

Vijeta Verma
School of Law
CHRIST University
Delhi, India

Muhammed Murat Yanaşoğlu
Department of AI, Data Analytics, and Software Engineering Departments
AI and IoT Research Center
AI and Informatics Faculty
Near East University
Mersin, Turkey

Zainab Babangida Umar
Department of AI, Data Analytics, and Software Engineering Departments
AI and IoT Research Center
AI and Informatics Faculty
Near East University
Mersin, Turkey

Matin Taghian Zadeh
Department of AI, Data Analytics, and Software Engineering Departments
AI and IoT Research Center
AI and Informatics Faculty
Near East University
Mersin, Turkey

Preface

This book explores the emerging landscape of sustainable unmanned aerial vehicle (UAV) ecosystems by combining rapidly evolving artificial intelligence, blockchain, and quantum technologies to redefine next-generation aerial systems and smart mobility. As editors, we are delighted to present how this collection addresses the growing demands for secure, intelligent, and energy-efficient UAV networks from leading researchers and practitioners in the field.

The volume starts with a wide range of topics, including post-quantum cryptography and security models, blockchain technology for decentralized cloud-based UAVs, clinical decision support systems, predictive intelligence, adaptive routing, bioinspired swarm intelligence algorithms, low-carbon and energy-aware UAV operations, and the integration of legal frameworks for sustainable UAV ecosystems. Subsequent chapters explore approaches to understanding complex UAV systems and highlight the transformative potential in addressing critical challenges such as scalability, security, energy efficiency, and real-time sustainable UAV networks.

We would like to express our gratitude to the contributors for their expertise and insights in this volume. Their dedication and commitment to advancing the field have made this book a valuable resource for researchers, practitioners, and students interested in the design of scalable, secure, and sustainable UAV ecosystems. We hope that this book serves as a catalyst for further exploration and innovation in the field of next-generation aerial sustainable ecosystems. We invite readers to delve into the chapters, engage with the ideas presented, and envision the transformative possibilities brought to the realm of intelligent and sustainable aerial systems.

Chapter 1

Cybersecurity foundations for sustainable UAV ecosystems

Pasula Sowjanya, Enugurthi Harika, J. Balaraju, Balusupati Veera Venkata Siva Prasad, and Manpinder Singh Panesar

1.1 INTRODUCTION TO CYBERSECURITY IN UAV ECOSYSTEMS

Drones, ground control stations, cloud platforms, and data analytics systems work together as a unit to enable unmanned aerial vehicle (UAV) ecosystems. Given that UAVs are typically used in areas such as smart cities, agriculture, disaster management, and logistics, they present an easy target for cyberattacks. The cybersecurity of UAV systems is intended to protect data, communications, and control systems, missions, and threats of attacks, such as hacking, spoofing, jamming, and unauthorized access. It requires an efficient cybersecurity foundation to offer secure, reliable, and sustainable UAV operations. It also creates confidence, compliance regulations, and sustainable deployment of UAV technologies in delicate and significant tasks [1].

1.1.1 Role of UAVs in sustainable and smart systems

UAVs are important for the creation of sustainable and intelligent systems since they possess efficient, data-driven, and adaptable solutions across numerous areas. Smart cities use UAVs to monitor traffic, conduct inspections on infrastructure, provide security to people, and conduct environmental surveillance, thereby reducing the number of human resources, time, and cost of operation. They are able to collect real-time, high-resolution data that can help the city planners to make good decisions on energy management, waste management, and urban development. UAVs have a significant application in the environmental sustainability sector in the areas of precision agriculture, forestry monitoring, wildlife protection, and climate research. They allow proper estimation of crop health, water and fertilizer use, and pest or disease outbreaks, thus reducing wastage of resources and environmental harm [2]. Another area of application of UAVs in disaster management is the provision of early warning, damage assessment, and search-and-rescue, which enhances response time and saves lives. UAVs are used in smart mobility and logistics to provide last-mile delivery, traffic

DOI: 10.1201/9781003731580-1

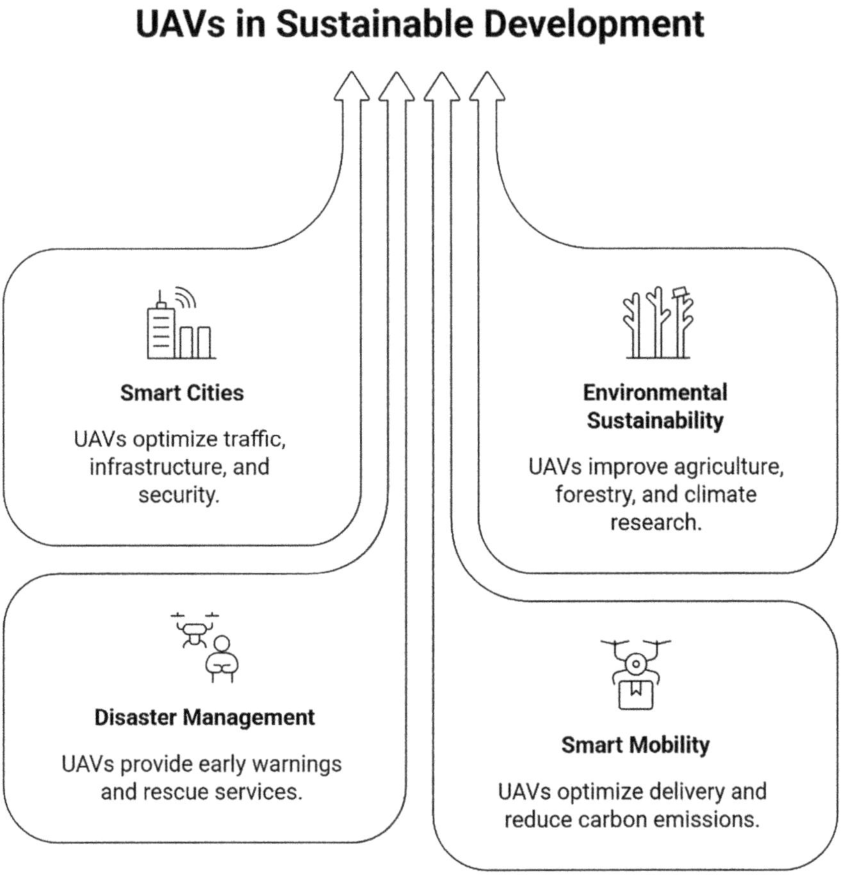

Figure 1.1 UAVs for smart sustainable development.

measurement, and optimization of routes and decrease fuel usage and carbon emissions. All in all, UAVs are smart sensing and communication devices that improve efficiency, resilience, and sustainability of contemporary smart systems, which are very much aligned with the objectives of sustainable development in the world. Figure 1.1 depicts the role of UAVs in smart sustainable development.

1.1.2 Cybersecurity as a pillar of UAV sustainability

Cybersecurity is one of the pillars in attaining sustainable UAV ecosystems. UAVs require safe communication connection, software, and data transfer in order to conduct safe and reliable operations. Failure to protect these systems may lead to cyberattacks, including hacking, spoofing, tampering of data, or denial-of-service which can result in missions being derailed and create severe safety threats. These failures may decrease the trust in UAV technologies and restrict their usage in the long-term. Effective cybersecurity practices can assist in maintaining the integrity of data, system availability, and privacy of users. Authentication, encryption, and access control are used to ensure that UAVs are not used, and unauthorized and sensitive information is not accessed. Cybersecurity also helps in regulatory compliance and responsible deployment, especially in sensitive sectors such as the health sector, transport, and defense. Cybersecurity assists in reducing the risks and increasing reliability to perform the UAV operations throughout effectively and in an environmentally friendly manner. Therefore, cybersecurity is not a technical requirement but a precondition of the further evolution and recognition of UAV systems in intelligent and forward-thinking space [3].

1.1.3 Objectives

The key objectives of this chapter are:

- To explain the importance of cybersecurity in ensuring safe and sustainable UAV ecosystems.
- To identify key cyber threats and vulnerabilities affecting UAV systems.
- To highlight essential security mechanisms and best practices for UAV operations.
- To emphasize the role of cybersecurity in building trust and long-term adoption of UAV technologies.

1.2 UAV SYSTEM ARCHITECTURE AND THREAT LANDSCAPE

A standard UAV system has a system architecture comprising the flying platform, a sensor onboard the platform, flight control, communication, a ground control station, and cloud or data processing services in the back end. The combination of these components assists in supporting navigation, data collection, command, and control. UAVs are also very dependent on wireless communication and software-controlled operations, which expose them to cyber risks. Attacks that are covered in the threat landscape comprise GPS spoofing, signal spoofing, malware injection, data interception, and unauthorized access to control systems. UAVs can also be stolen physically, resulting in data theft or manipulations of the system. Moreover, the security of cloud systems and third-party software can be compromised, potentially compromising the entire UAV system. To design effective security strategies, it is important to know the system architecture as well as the threat landscape related to the system. Such a clear perspective on how the threats attack various elements can be useful in the application of appropriate protection measures and the maintenance of safe and robust UAV operations [4].

1.2.1 UAV components: air, ground, and communication segments

The UAV system often consists of three primary parts, including the air component, the ground component, and the communication component. The air segment comprises the unmanned aircraft itself, onboard hardware, and software, including sensors, cameras, navigation systems, flight controllers, and embedded processors. This component deals with data acquisition, airplane stability, and flight control. Ground segment comprises ground control stations, operator interfaces, monitoring units, and data storage units. It enables human operators to plan missions, send control commands, receive telemetry, and analyze data collected. Ground stations can be both fixed and mobile and usually are connected to cloud platforms to process data and provide decision support. The communication segment relates the air segment and the ground segment via wireless networks like radio frequency, satellite, cellular, or 5G. It enables command and control, real-time data delivery, and system updates. As this segment contains very vital information, it is very susceptible to cyber threats such as interception, jamming, and spoofing. All three segments must be integrated safely and securely to achieve reliable and safe UAV operations [5]. Various UAV components are summarized in Table 1.1.

Table 1.1 UAV components, functions, and cybersecurity aspects

UAV segment	*Key components*	*Primary functions*	*Cybersecurity concerns*
Air segment	Unmanned aircraft, sensors, cameras, navigation systems, flight controllers, and embedded processors	Data acquisition, flight stability, navigation, and autonomous control	Malware injection, sensor spoofing, and unauthorized firmware access
Ground segment	Ground control stations, operator interfaces, monitoring units, data storage, and cloud platforms	Mission planning, command transmission, telemetry reception, and data analysis	Unauthorized access, data breaches, and insider threats
Communication segment	RF links, satellite communication, and cellular/5G networks	Command and control, real-time data transmission, and system updates	Interception, jamming, spoofing, and denial-of-service attacks

1.2.2 Attack surfaces in UAV ecosystems

Attack surfaces in the UAV ecosystem are all the points where an attacker can attempt to gain unauthorized access or interfere with the functioning of the system. The communication between the UAV and the ground control station is one of the key attack surfaces. Being based on wireless transmission, it is susceptible to eavesdropping, jamming, spoofing, and man-in-the-middle attacks. Malware or malicious code can be injected into the UAV's onboard software and firmware, which is another critical area of attack. Attack surfaces are also high in ground control stations. Weak authentication, outdated software, or ineffective access controls may enable attackers to hijack UAV operations or intercept sensitive mission data. Another factor is the use of cloud services and backend servers to store information and conduct analytics, which contributes to the growth of the attack surface, especially when third-party services are involved. Further, physical access can be used to tamper with hardware and steal data from UAVs, such as during the maintenance process or during forced landings. These attack surfaces are significant to understanding how to apply successful cybersecurity and deliver secure, resilient, and sustainable UAV ecosystems [6]. The attack surface in UAV systems is presented in Figure 1.2.

1.2.3 Cyber-physical security risks

Cyber-physical security threats arise when cyberattacks have a direct impact on the physical behavior and security of UAV systems. Since UAVs are

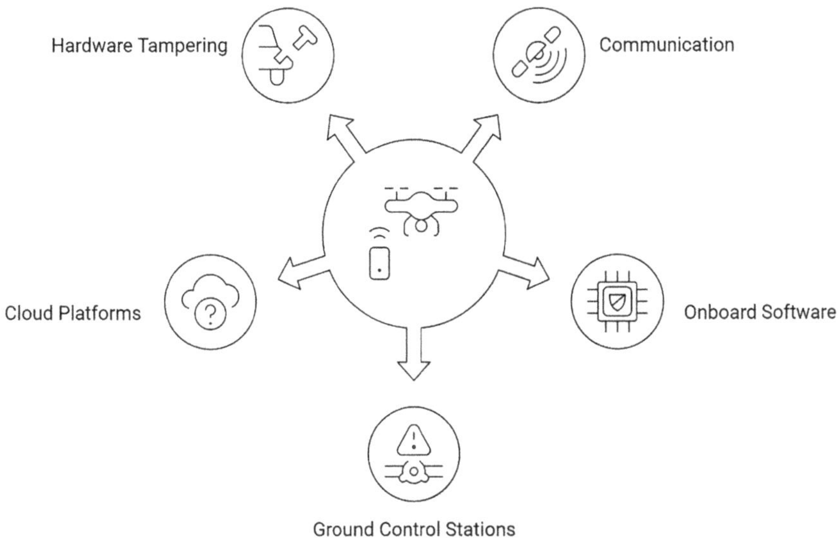

Figure 1.2 UAV systems-attack surface.

directly related to each other in terms of software, sensors, communication systems, and mechanical parts, any cyberattack can easily transform into physical damage or unstable functioning. For example, GPS spoofing can provide a UAV with incorrect position information, resulting in navigation errors or accidents. On the same note, hijacking of control or jamming of signals can result in loss of control, forced landing, or mission failure. Onboard sensors or flight control software can be manipulated, and this may result in wrong decision-making, unbalanced flight, or collision with obstacles. Attacks on ground control stations or cloud platform attacks can alter the flight parameters of missions, disclose confidential data, and even leave entire groups of UAVs out of commission. These risks are especially high in systems such as disaster response, medical care delivery, inspection of the key infrastructure, and intelligent transportation, when the safety and reliability of the systems are the highest priority. Other risks that are brought about by the cyber-physical attacks include public safety, privacy, and the environment. Unethical surveillance, information abuse, or physical destruction through hacked UAVs can erode the amount of trust the masses have and slow the pace of technology adoption. The mitigation of cyber-physical security threats is therefore a collective security approach that considers cybersecurity, physical security, and systems resilience to ensure safe and sustainable UAV operation [7].

1.3 COMMON CYBER THREATS AND VULNERABILITIES IN UAVS

UAVs are becoming increasingly susceptible to cyber insecurity because they are wireless, software-based, and interconnected systems. Common cyber-attacks include GPS spoofing, signal jamming, data interception, malware injection, and unauthorized access to control systems. The vulnerabilities can include weak authentication, the absence of encryption for communication connections, the insecurity of the firmware updates, and the absence of onboard security resources. Cyberattacks may also be employed against ground control stations and cloud platforms that are not well secured. Such risks can lead to the loss of control, unprotected data, failure in the mission, and threats to safety. Knowing the general cyber threats and vulnerabilities will be critical in ensuring effective system security and reliability, safety, and sustainability of UAV operation [8].

1.3.1 GPS spoofing and signal jamming

Other severe threats to cybersecurity that have a direct impact on a UAV's navigation and control include GPS spoofing and signal jamming. GPS spoofing is a form of cyberattack in which an attacker sends inappropriate

GPS signals to confuse a UAV regarding its true position, velocity, or time. Consequently, the UAV can fail to follow its scheduled path, fly in restricted space, or even crash down. Signal jamming, on the other hand, is the interruption or saturation of legitimate communication or GPS signals through strong interference. This may result in the loss of navigation data or interfere with the command-and-control connection between the UAV and the ground station. These two attacks may result in mission failure, an unsafe landing, or total loss of the UAV. These threats are especially harmful in critical applications such as disaster response, surveillance, and delivery services. GPS spoofing and jamming of UAVs should be counteracted to enable safe, reliable, and sustainable operations.

1.3.2 Malware, firmware, and software attacks

UAVs rely heavily on software and firmware to manage their flight, navigation, sensors, communication, and data processing. This dependency puts them at risk of malware and software-based attacks that could lead to vulnerability of integrity, availability, and safety of operations. Malware attacks involve the introduction of malicious code into the UAV systems through malicious files, hacked networks, or malicious updates. When malware is activated, UAVs can be controlled by outside parties, confidential data can be stolen, and even the operations of UAVs can be stopped. Firmware attacks are those attacks that target low-level software embedded in UAV hardware. Important processes such as sensors, motors, and communication modules are controlled by firmware, and thus any hacking can have long-term impacts on the performance of the UAV. The vulnerabilities in firmware that are either old or not updated can be exploit by hackers to gain unauthorized access or install persistent malware that remains active even after system reboots. Software attacks are also available, and they are founded on ground control station program vulnerabilities, mission planning program vulnerabilities, or cloud-based system vulnerabilities. The hijacking of UAV operations, manipulation of mission information, or compromise of multiple UAVs simultaneously can be achieved through weak authentication, insecure APIs, or unencrypted communication channels. Good security practices that can reduce these risks include regular updates of software and firmware, verification of the code integrity, use of secure communication protocols, malware detection, and strong access controls. The UAV software and firmware security are crucial to the delivery of safe, reliable, and sustainable UAV operations within sensitive and high-stakes environments [9].

1.3.3 Data breaches and privacy violations

UAVs produce and broadcast huge amounts of data, such as images, videos, sensor measurements, position data, and logs of operations. This information is usually stored either in onboard computing devices, ground stations, or cloud computing platforms to be analyzed and used to make decisions. Nevertheless, collection, storage, and transmission of sensitive information put UAV ecosystems at risk of data breaches and privacy breaches. Poor authentication, unsecured communication connections, unencrypted data storage, or anything that takes advantage of software vulnerabilities can lead to unauthorized access to this data. The personal, organizational, or mission-critical information might leak out due to data breaches in UAV systems. For example, aerial surveillance can record personal properties or critical infrastructures, which, in the hands of unscrupulous individuals, can be abused to carry out espionage, theft, or other criminal offenses. Ethical and legal issues are also related to the privacy intrusions in cases when the UAV operations are operated without appropriate permission or legal regulations. To guarantee privacy and security of data, it is necessary to ensure that communication and storage are encrypted with strong encryption tools, authentication is performed through effective methods, access is controlled strictly and legal frameworks are adhered to. Also, misuse of information obtained can be averted through anonymization and responsible data handling practices. Securing the UAV data is crucial in ensuring that the people have confidence, the regulations are upheld and that the use of UAV technologies is sustainable.

1.3.4 Insider and supply-chain threats

The UAV ecosystems are prone to external cyberattacks as well as insider and supply-chain attacks. Insider threats are inflicted when employees, contractors, or operators misuse their access to UAV systems either intentionally or accidentally, resulting in a leakage of data, system interference, or unauthorized alterations of missions. The nature of these threats is especially risky since insiders already know system architecture, operations, and security protocols. Supply-chain threats are when parts, programs, or services of third-party vendors are hacked prior to being incorporated into UAV systems. It may be introduced at any point, whether it is during the manufacturing of the hardware or in the software updates. These vulnerabilities may be used in the future to obtain unauthorized access, control data, or impair the performance of the system. To reduce insider and supply-chain risks, stringent access control, frequent audits, employee training, verification of vendors, and safe procurement methods are to be employed. To maintain integrity, reliability, and long-term sustainability of the UAV operations in sensitive and critical applications, it is necessary to address these threats.

Common cyber threats and vulnerabilities in UAV systems are summarized in Table 1.2.

Table 1.2 UAV systems—threats and vulnerabilities

Threat category	*Description*	*Key vulnerabilities exploited*	*Potential impacts*	*Mitigation measures*
General cyber threats	UAVs are exposed to cyber risks due to wireless links, software dependence, and system interconnectivity	Weak authentication, lack of encryption, insecure firmware updates, and limited onboard security	Loss of control, mission failure, data breach, safety hazards	Strong authentication, encrypted communication, secure updates, and layered security
GPS spoofing and signal jamming	Fake GPS signals mislead UAV navigation; jamming disrupts GPS or command links	Unprotected GPS signals, weak anti-jamming mechanisms	Navigation failure, unsafe landing, UAV loss, restricted airspace violation	Anti-spoofing techniques, redundant navigation, and signal monitoring
Malware, firmware and software attacks	Malicious code targets UAV software, firmware, or ground/cloud systems	Outdated firmware, insecure APIs, unpatched software, and weak access control	Hijacked UAVs, data theft, and persistent system compromise	Secure boot, code integrity checks, regular updates, and malware detection
Data breaches and privacy violations	Unauthorized access to images, videos, sensors, and mission data	Unencrypted storage, poor access control, and insecure data transmission	Privacy invasion, espionage, legal, and ethical violations	Data encryption, strict access control, anonymization, and legal compliance
Insider and supply-chain threats	Misuse of privileged access or compromised third-party hardware/software	Excessive privileges, lack of audits, and unverified vendors	Data leakage, system manipulation, and long-term hidden vulnerabilities	Role-based access, audits, staff training, and trusted procurement

1.4 SECURE COMMUNICATION AND DATA PROTECTION MECHANISMS

Protecting information and ensuring secure communication are critical to the operation of the UAVs. UAVs are dependent on wireless connections to send commands, telemetry, and mission data; thus, they are prone to being intercepted, altered, or accessed by individuals without authorization. Encryption, secure authentication, and integrity checks are implemented to ensure that UAV systems are accessed or controlled only by authorized users. UAVs contain sensitive information protected by data protection measures, such as secure storage, access controls and anonymization. These solutions not only help avoid cyberattacks but also facilitate regulatory compliance, privacy, and trust. Resilient and sustainable UAV ecosystems are based on strong communication and data security [10].

1.4.1 Authentication and access control

Authentication and access control cannot be discussed outside the framework of UAV cybersecurity, as they ensure that access to UAV operations and information is limited to authorized users and systems. Authentication determines the identities of operators, devices, and applications, as those identities allow access to the system. Examples of such common options include passwords, PINs, biometric authentication, digital certificates, and multi-factor authentication (MFA). Strong authentication enables control by the authorized bodies, and it reduces the possibilities of hijacking, manipulation of missions, or theft of information. One of the attributes of what the authenticated users or systems can do in the UAV ecosystem is access control. The authorization of access to sensitive functions, flight commands, or data is determined by role, responsibility, or attributes through role-based access control (RBAC), attribute-based access control (ABAC), and least-privilege principles, respectively. Potential damage of insider threats, hacked accounts, malware, etc. is blocked by access control. Authentication and access control are both part of the first layer of protection against UAV security threats to communication links, ground stations, cloud-based systems, and onboard systems. They are required for the long-run viability of the UAV missions, data secrecy, and integrity of their operations [11].

1.4.2 Encryption and key management for UAV networks

The management and encryption are the most critical issues in protecting the communications and data of UAVs. The data being conveyed by UAV networks are of a sensitive nature, including telemetry, control commands, and missions that are being transferred through wireless media whose data are easily intercepted and altered. Encryption converts this information to unreadable forms by unauthorized users, which ensures confidentiality and integrity. The resources are secured in an efficient and fast manner by symmetric encryption and key exchange and authentication by asymmetric encryption. Key management is employed in making sure that the process of creating encryption keys, their distribution, storage, rotation, and revocation is in a way that is secure. The failure to handle the keys appropriately may render the encryption useless, putting the UAV systems at risk of cyberattacks. Some of the techniques applicable toward ensuring the strength and reliability of encryption are public key infrastructure (PKI), secure key storage, and automated key rotation. The use of the strong encryption with the key management will protect the UAV communication links and the stored information, eliminate the unauthorized access, and maintain the integrity of the mission. These are critical toward creating resilient, trusted, and sustainable UAV ecosystems.

1.4.3 Secure command, control, and telemetry links

UAV activities are based on links of command, control, and telemetry (C2T), which is a communication between the UAV, ground control stations, and cloud platforms. These connections are flight commands, mission information, and real-time telemetry and are hence the most susceptible to cyberattacks such as interception, spoofing, jamming, and unauthorized access. Poor C2T links can cause failure in the mission, a hijack of UAVs, or unsafe operations, particularly in cases where the system is required in emergency conditions like disaster management, surveillance, and health service provision. The securing of C2T links has many layers of protection associated with it. Encryption ensures that the data transmitted is confidential and cannot be deciphered by the attackers. The authentication systems are used to ensure that the commands are given by the authorized operators or systems. Checks on integrity detect the tampering or alteration of a message in its transmission. Frequency hopping and anti-jamming techniques and communication channel redundancy can also increase resilience against jamming or denial-of-service attacks. Moreover, the secure C2T connections must be effective even in the case of resource limitations, like in the case of the UAVs, low processing units, and battery duration. These security measures lead to safe, reliable, and trustworthy UAV operations, which can be effectively integrated into resilient and sustainable UAV systems that can be safely integrated within smart and critical infrastructure systems.

1.5 BLOCKCHAIN AND DISTRIBUTED SECURITY FOR UAV NETWORKS

Blockchain and distributed security is a good field that can be used to guarantee the security of the UAV networks against the cyber threat. Blockchain gives the possibility to make the data, commands, and transactions tamper-proof, traceable, and verifiable without the interference of any central authority with the help of decentralized ledgers. This may particularly be handy to UAV ecosystems, whereby several drones, ground stations, and cloud platforms are communicating in dynamic scenarios. Distributed security systems (consensus and smart contracts) enhance trust, prevent unauthorized access, and reduce the risk of manipulating information. Applications of blockchain make the UAV tasks more resilient, transparent, and sustainable in the critical and multi-agent tasks [11–14].

1.5.1 Decentralized trust models for UAV ecosystems

Decentralized trust models provide the safe method of interaction between UAVs and ground stations and cloud platforms without the assistance of a single central authority. The traditional UAV networks with a central controller or a server have the tendency of making the network a matter of trust, thus making it a point of failure and vulnerable to attacks. Decentralized models address this issue by distributing trust among a large number of nodes in a number of ways, such as blockchain, peer-to-peer verification, or consensus protocols. A decentralized UAV ecosystem enables the exchange of mission data and flight plans and operational commands among multiple drones and stakeholders in a secure manner. The network checks every transaction or data transfer, and it is guaranteed integrity, authenticity, and transparency. Smart contracts can automatically authenticate, enforce policies, and access controls. The models also increase the resilience of the system to attack, as the failure of one node does not affect the entire system. Especially significant in the connection with decentralized trust are multi-UAV operations, swarm systems, and collaborative missions of UAVs that autonomously interact with third-party services. By reducing reliance on centralized authorities, these models contribute to greater reliability, transparency, and security and help to make the UAV technology sustainable in sensitive operations, including logistics, surveillance, environmental monitoring, and disaster management.

1.5.2 Blockchain-based identity and data integrity

The identity, security, and data integrity in the UAV ecosystem can be solved through blockchain technology. Each UAV, operator, or device can be issued a blockchain-based digital identity that is verified, tamper-resistant, and resistant to impersonation. This prevents unauthorized access and ensures that only trusted parties are involved in UAV operations. Flight logs, sensor readings, or mission reports that UAVs distribute or send can also be stored on a blockchain. The records are encrypted to the final record and hence cannot be changed or modified. It is easy to detect any manipulation or interference of information, and this gives credibility and authenticity to the information. Blockchain-based systems for identity and data integrity can also be applied to enhance transparency, accountability, and regulatory compliance. These capabilities enable safe and trackable drone ground station and cloud communications in multi-UAV, swarm, or team settings. The blockchain improves the reliability, credibility, and sustainability of UAV ecosystems by combining identity management and unalterable data entries.

1.5.3 Smart contracts for secure UAV operations

Smart contracts are computer programs that may be available on a blockchain and automatically implement the set rules and conditions without human intervention. They serve as a safe operations management tool within the UAV ecosystem to ensure that missions, data exchange, and transactions adhere to accepted protocols. For example, a smart contract can be deployed to allow flight plans when all the safety inspection and regulatory prerequisites are satisfied, or it can be deployed to release mission data to authorized entities in controlled conditions. Automation of verification and enforcement reduces the likelihood of human error, unauthorized access, and malicious manipulation of smart contracts. They also enable safe collaboration among different UAVs, ground stations, and service providers in multi-agent systems or in swarm systems. Each transaction fulfilled with the help of smart contracts is recorded on the blockchain and is transparent, traceable, and accountable. The introduction of smart contracts into the UAV business would enhance the security, efficiency in business operations, and trust in autonomous and semi-autonomous flights. The resilience of transparent and sustainable UAV ecosystems can be supported by the methodology in areas such as logistics, surveillance, disaster management, and urban mobility, where the safety and reliability of operations are paramount.

1.6 AI-DRIVEN CYBERSECURITY FOR UAV SYSTEMS

Artificial intelligence and machine learning in AI-based cybersecurity are used to detect, prevent, and respond to cyber threats in UAV systems. UAVs operate in dynamic environments and generate enormous volumes of data, which is why the implementation of conventional security solutions is insufficient. Using network traffic, flight behavior, and sensor data, AI algorithms can identify anomalies, anticipate potential attacks, and automatically neutralize the threat in real time. AI can enhance system resilience, reduce human interactions, and accelerate response times through continuous learning based on new threats. The use of AI-based cybersecurity makes the UAV operations safe, reliable, and sustainable in the implementation of critical and complex operations [15].

1.6.1 Intrusion detection and anomaly detection using AI

Intrusion detection and anomaly detection are significant in the protection of UAV systems against cyber threats. Traditional security measures are not normally capable of sustaining the dynamic setting, where UAVs are utilized. The machine learning (ML) and deep learning (DL) procedures of AI-based solutions offer more advanced solutions for suspicious activity since they examine and process network traffic, sensor information, and system actions automatically. The artificial intelligence (AI-based) intrusion detection system (IDS) can detect real-time unauthorized access, malware, and command injection attacks. These systems have the ability to change according to the changing threats as they are self-educated on the pattern of the attacks in the past, and they provide proactive protection. Anomaly detection, in turn, is interested in identifying the anomalies in the regular behavior of UAVs, such as unusual routes of flight, abnormal communication patterns, or unusual sensor measurements. These deviations can be detected early, enabling AI systems to prevent potential cyber-physical damage or mission failure. AI-based detection methods can also offer threat prioritization, false alarm reduction, and automated response, which is especially important because of the lack of human control in UAV missions. With the addition of AI-based intrusion and anomaly detection and other cybersecurity services like encryption, authentication, and secure communication, the resilience, reliability, and sustainability of UAV ecosystems can be dramatically improved to enable safe deployment in harsh and high-risk conditions [16].

1.6.2 Machine learning for threat prediction and response

Machine learning (ML) has emerged as a game-changer in improving the cybersecurity of the UAV systems through predictive threat detection and automatic response systems. UAVs have to operate in dynamic, complex environments that are in permanent interaction with one another, with ground stations and cloud platforms, and with other UAVs. Conventional rule-based security solutions are not capable of addressing fast-changing cyber threats such as GPS spoofing, signal jamming, malware attacks, and insider threats. However, ML algorithms can process large volumes of network, sensor, and operational data to detect patterns and correlations indicative of new threats. Predictive threat detection uses both supervised and unsupervised ML models to estimate the likelihood of an attack before it occurs. As an example, anomaly detection models can learn normal UAV behavior, including flight paths, communication patterns, and sensor readings, and indicate abnormal behavior that might flag malicious activity. The incoming network traffic can also be classified as safe or malicious by the classification models, allowing operators to prioritize critical alerts. In addition to detection, ML may be used to assist with automated response plans. Adaptive algorithms and reinforcement learning can be used to enable UAV systems to change flight paths, communication routes, or access controls in real time in response to a threat, minimize operational disturbances, and increase safety. ML can also keep refining its forecasts by training on novel attack strategies, making UAV networks more resistant to advanced attackers. By combining threat prediction and ML-based response with encryption, secure communication, and access control, UAV ecosystems achieve proactive defense, operational reliability, and sustainable cybersecurity. It is a necessary feature of applications in the disaster management, urban logistics, environmental monitoring, and defense domains, where safety and mission continuity are paramount [17].

1.6.3 Explainable AI for trustworthy UAV security

Explainable AI (XAI) is a tool that can improve the cybersecurity of UAVs by making AI decisions understandable, transparent, and accountable. Although machine learning models can spot threats, anticipate attacks, and provide automated responses, their complex algorithms often make them black boxes, so operators cannot rely on their results. XAI solves this by providing a clear explanation of how decisions are made, including which data, patterns, or anomalies were used to make a prediction or take an action. XAI is used in UAV systems to make operators know why a flight deviation was alerted, why some commands were blocked, or why a potential threat was prioritized. Such transparency enhances trust in independent decisions and allows working in closer collaboration with AI. Regulatory compliance, accountability, and forensic analysis following the incident are also supported by XAI to ensure that UAV operations are safe, ethical, and reliable. XAI supports trustful, robust, and sustainable cybersecurity for UAV ecosystems by integrating the predictive capabilities of AI with the ability to interpret its results [18].

Various AI-driven cybersecurity mechanisms for UAV systems are presented in Table 1.3.

1.7 REGULATORY, ETHICAL, AND PRIVACY CONSIDERATIONS

UAV cybersecurity is a technical problem that is accompanied by regulatory, ethical, and privacy issues. Safety standards of UAVs, data security, and airspace regulations are established by regulations, which maintain accountability and the safety of the population. Responsible UAVs address ethical issues such as preventing illegal surveillance, misuse of data, or injury to individuals and society. Privacy is of paramount concern since UAVs tend to record sensitive

Table 1.3 Various AI-driven cybersecurity mechanisms for UAV systems

Aspect	*Focus area*	*AI techniques used*	*Key functions*	*Benefits for UAV security*
AI-driven cybersecurity for UAVs	Intelligent cyber defense for dynamic UAV environments	AI, ML, and DL	Real-time threat detection, prediction, and mitigation	Improved resilience, faster response, reduced human intervention, and sustainable operations
Intrusion and anomaly detection	Detecting abnormal behavior and cyber intrusions	ML-based IDS, deep learning, and anomaly detection models	Identify unauthorized access, malware, abnormal flight paths, and unusual traffic	Early threat detection, fewer false alarms, and enhanced reliability
ML for threat prediction and response	Proactive and automated cybersecurity	Supervised and unsupervised ML and reinforcement learning	Predict attacks, classify traffic, adapt routes, and access control	Proactive defense, minimized mission disruption, and improved safety
Explainable AI (XAI)	Transparent and trustworthy AI decisions	XAI models and interpretable ML	Explain alerts, decisions, and automated actions	Increased trust, regulatory compliance, accountability, and ethical security

personal, commercial, or environmental data. The consideration of such concerns, as well as technical security, will make sure that the UAV activities will be legally legitimate, ethically accountable, and privacy-conscious to ensure the building of trust in the population and the environmentally friendly and safe application of UAV technologies.

1.7.1 UAV cybersecurity standards and compliance

UAV cybersecurity and compliance provide a systematic framework to ensure safe, reliable, and secure operations. The best practices, technical requirements, and guidelines for protecting UAV hardware, software, communication links, and data against cyber threats are defined by standards. There are guidelines on cybersecurity, risk management, and system safety for UAV operations provided by organizations such as the International Organization for Standardization (ISO), the Institute of Electrical and Electronics Engineers (IEEE), and aviation authorities. These standards play a crucial role in eliminating risks like unauthorized access, data breaches, signal interference, and cyber-physical attacks. It also provides interoperability, operational stability, and regulatory compliance across a wide range of UAV platforms and applications. UAV systems are certified and audited to ensure they meet security standards, including encryption, authentication, access control, and secure software updates. By making UAV activities conform to the accepted cybersecurity standards, operators improve the safety level, preserve the confidence of the population, and minimize the legal and operational risks. The sustainable implementation of UAV technologies in commercial, industrial, and critical infrastructure applications requires standardized compliance [19].

1.7.2 Ethical use and responsible security design

In UAV cybersecurity, ethical use and responsible security design are vital values that should be adopted so that UAV technologies are implemented safely, fairly, and sustainably. Ethical use focuses on the consideration of human rights, privacy, and societal standards accompanied by the avoidance of using UAVs to illegally spy, misuse the information, or cause damage. To ensure that people trust the operators, the operators should operate under clear policies, seek informed consent where necessary, and abide by legal and regulatory frameworks. Responsible security design involves integrating cybersecurity considerations throughout the UAV lifecycle, including design and production, operation, and maintenance. These involve the use of effective authentication, encryption, access control, software updates, and anomaly detection systems. The physical elements of the UAV, communication connections, and cloud platforms should also be considered in terms of security to ensure it is resistant to cyber-physical attacks. Ethical application together with responsible design will create UAV systems that are secure, socially and environmentally responsible. It promotes responsibility, minimizes the risk of abuse, and encourages adherence to rules and regulations. By prioritizing these principles, the UAV operators and developers will be able to create sustainable environments that would balance the technological innovation with the safety, privacy, and welfare of society [20].

1.7.3 Privacy protection in civil and commercial UAV applications

Privacy protection is a key issue in civil and business UAV use, since drones can often collect sensitive data, such as photos, videos, geolocation data, and operational records. UAVs are used in areas like logistics, agriculture, infrastructure inspection, and urban surveillance to record precise environmental and personal data, which, when used inappropriately or carelessly stored, may result in privacy invasion and legal consequences. Protection of privacy entails the adoption of sound technical and procedural practices. Sensitive information cannot be accessed by unauthorized individuals because of data encryption, secure storage, and access control. Individual identities can be safeguarded through anonymization and aggregation means, and relevant analysis of data collected can be conducted. Privacy rules and regulations, including the General Data Protection Regulations (GDPR) or the regulations of local aviation authorities, should be adhered to ensure the rights of the users and keep the population's trust. The operators should also formulate clear policies in relation to data collection, retention, and sharing, whereby the UAV missions should be transparent and consent-based. Monitoring and auditing of an action are conducted continuously to identify privacy risks and impose accountability. With privacy protection incorporated into the UAV design, operation, and governance, civil and commercial applications can be safely, ethically, and sustainably utilized, balancing innovation and responsible data management.

1.8 CYBERSECURITY FOR SUSTAINABLE AND SMART UAV APPLICATIONS

Sustainable and smart UAV applications are impossible without cybersecurity as a crucial enabler. Smart cities, precision agriculture, disaster management, logistics, and environmental monitoring are fields where UAVs are gaining significance; in these areas, the dependability and safety of operations are crucial. The threat of cyberattacks like hacking, data breaches, GPS spoofing, and signal jamming may affect safety, privacy, and mission success. Such a strong cybersecurity approach, including secure communication, encryption, access control, AI-based threat detection, and blockchain-based integrity assurance, ensures operational stability, regulatory compliance, and public trust. Good cybersecurity facilitates the sustainable integration of UAVs into smart, networked, and robust systems [21, 22].

1.8.1 Smart mobility and urban airspace security

UAVs are becoming an integral part of intelligent mobility, including urban deliveries, traffic monitoring, emergency response, and airspace control. Their usage in areas with high population density creates its own cybersecurity issues, as UAVs interact with a large number of drones, ground stations, and IoT-based infrastructure. GPS spoofing, signal jamming, unauthorized access, and data leaks are threats that may disrupt operations, undermine safety, and jeopardize public trust. To ensure the safety of the airspace over large cities, it is necessary to implement robust security measures such as encrypted communications, secure authentication, real-time intrusion detection, and AI-driven anomaly monitoring. Safety is also enhanced by regulatory compliance, flight path verification, and collision-avoidance procedures. Technical security, coupled with operational governance, can enable UAVs to support smart mobility efficiently, safely, and sustainably, thereby making them a reliable part of an urban transport and airspace management system.

1.8.2 Environmental monitoring and critical infrastructure protection

UAVs have been a significant part of environmental monitoring and safeguarding critical infrastructure, providing effective, efficient, and real-time data collection services. They are applied in air and water quality monitoring; deforestation monitoring; natural disaster monitoring; and energy, transportation, and industrial facilities monitoring. Nevertheless, the weakness and sensitivity of such applications make UAVs the best targets for cyberattacks. Signal jamming, GPS spoofing, malware injection, and data breaches are among the threats that can interfere with missions, compromise gathered information, or even cause harm to critical systems. Some measures to ensure cybersecurity in such settings include secure communication channels, encrypted data transfer, strong authentication, and AI-driven anomaly detection. Blockchain-based intelligence checks can ensure data is not tampered with, while access controls and regulatory compliance will guarantee sensitive data is not compromised. With robust cybersecurity, UAVs will be safe and reliable, enabling sustainable environmental surveillance, critical infrastructure protection, and preventing people from losing confidence in high-stakes and sensitive missions.

1.8.3 Disaster response and public safety UAV security

UAVs are currently used in disaster management and civic safety operations, such as search-and-rescue, wildfire surveillance, flood management, and emergency medical services. These missions are usually unpredictable and hostile, and hence, cybersecurity is important to support secure and reliable operations. Missions can be disrupted, emergency response can be slowed, or lives can be endangered by cyberattacks, including signal jamming, GPS spoofing, unauthorized access, or data manipulation. In such situations, the protection of UAVs is to be of a multi-layered nature, which implies encrypted communications, high-level authentication, protective command-and-control connections, and AI-based threat detection. To ensure the security of sensitive victim and location information, access control and privacy measures are required. Unnecessary communication paths and anomaly detection increase system resilience during the most crucial tasks. Through sound cybersecurity, UAVs will be able to work safely, preserve the integrity of the mission, and assist in the effective response to disasters and community safety, which will provide timely interventions and protect communities during emergencies [23].

1.9 CHALLENGES AND FUTURE RESEARCH DIRECTIONS

Nevertheless, there are many cybersecurity challenges in UAV ecosystems despite great progress. UAVs deal with heterogeneous, dynamic, and resource-constrained environments, and it is hard to ensure robust security without compromising performance or flight time. Wireless communication links can be jammed, spoofed, and intercepted, and onboard software, firmware, and cloud platforms are susceptible to malware and data breaches. The additional challenges to securing UAV deployment are insider threats, supply chain vulnerabilities, and regulatory gaps. Also, the increasing use of AI, IoT, and blockchain brings complexity that requires explainable, reliable, and interoperable security solutions. Civil and commercial applications also require special consideration of privacy and ethical considerations. Future directions of the research include the creation of lightweight, adaptive, and autonomous security mechanisms that are appropriate to the constraints of UAVs. Proactive cybersecurity, enabled by AI-based threat prediction, anomaly detection, and explainable AI models, is in the pipeline. Decentralized structures using blockchain can enhance trust, identity, and data integrity. The exploration in the field of secure multi-UAV coordination, robust communications protocols, and privacy-sensitive methods can be identified. The standardization, regulatory, and ethical research will facilitate safe, responsible, and sustainable UAV operations. The development of these areas will make it possible to have robust, intelligent, and reliable UAV ecosystems to support smart, critical, and high-stakes applications [1, 3, 15, 22, 24].

Table 1.4 presents a summarization of key challenges and future research directions in the UAV ecosystem.

Table 1.4 Key challenges and future research directions in the UAV ecosystem

Segment	*Aspect*	*Description/focus*	*Impact/expected outcome*
Challenges	Resource constraints	Limited onboard power, computation, and memory	Security–performance trade-offs and reduced flight endurance
	Wireless communication risks	Jamming, spoofing, and interception of links	Loss of control, navigation errors, and mission failure
	Software and firmware vulnerabilities	Malware, insecure updates, and cloud weaknesses	System compromise and unreliable operations
	Insider and supply-chain threats	Misuse of access and compromised third-party components	Persistent hidden vulnerabilities
	System heterogeneity	Diverse hardware, software, and protocols	Difficulty in unified and scalable security
	Privacy and regulatory gaps	Inadequate civil and commercial UAV policies	Legal risks and loss of public trust
	Lightweight security mechanisms	Energy-efficient and adaptive protection methods	Strong security without reducing flight time
Future research directions	AI-driven proactive defense	Threat prediction, anomaly detection, and explainable AI	Early attack detection and trustworthy automation
	Blockchain-based security	Decentralized identity, trust, and data integrity	Improved transparency and tamper resistance
	Secure multi-UAV coordination	Cooperative and swarm-level security	Safe and scalable multi-UAV operations
	Robust communication protocols	Anti-jamming and resilient communication schemes	Reliable command and control

1.10 CONCLUSION

The safe, reliable, and sustainable implementation of the UAV ecosystems is enabled by cybersecurity. With the integration of UAVs into smart cities, environmental management, disaster management, logistics systems, and critical infrastructure, the threats to UAV security are diverse, including GPS spoofing, signal jamming, malware attacks, information breaches, and insider or supply chain attacks. It is necessary to secure all the UAV components, communication channels, and cloud platforms to ensure that mission integrity, citizens, and confidential information are not compromised. State-of-the-art technologies such as AI-based threat detection, blockchain-based identity and data integrity, and smart contracts provide new solutions for resilient and reliable UAV operations. Responsible deployment is also supported by regulatory compliance, privacy protection, and ethical issues. Nevertheless, with a solid cybersecurity implementation in place alongside operational and technological best practices, trust can be built, performance will be improved, and long-term adoption will be encouraged. Finally, an effective cybersecurity foundation is essential to achieving the full potential of UAVs for sustainable, smart, and critical use, ensuring safe, efficient, and socially responsible aerial operations.

REFERENCES

[1] R. Morshedi and S. Mojtaba Matinkhah, "Cybersecurity challenges and solutions in unmanned aerial vehicles (UAVs)," *Journal of Field Robotics*, vol. 43, no. 1, pp. 314–329, 2026.
[2] M. Aloqaily, O. Bouachir, I. Al Ridhawi, and A. Tzes, "An adaptive UAV positioning model for sustainable smart transportation," *Sustainable Cities and Society*, vol. 78, p. 103617, 2022.
[3] H. Alqahtani and G. Kumar, "Cybersecurity in electric and flying vehicles: Threats, challenges, AI solutions & future directions," *ACM Computing Surveys*, vol. 57, no. 4, pp. 1–34, 2024.
[4] K. A. Gromada and W. M. Stecz, "Designing a reliable UAV architecture operating in a real environment," *Applied Sciences*, vol. 12, no. 1, p. 294, 2021.
[5] C. Yan, L. Fu, J. Zhang, and J. Wang, "A comprehensive survey on UAV communication channel modeling," *IEEE Access*, vol. 7, pp. 107769–107792, 2019.
[6] O. Ceviz, S. Sen, and P. Sadioglu, "A survey of security in UAVs and FANETs: Issues, threats, analysis of attacks, and solutions," *IEEE Communications Surveys & Tutorials*, vol. 27, no. 5, pp. 3227–3565, 2025.
[7] C. Rani, H. Modares, R. Sriram, D. Mikulski, and F. L. Lewis, "Security of unmanned aerial vehicle systems against cyber-physical attacks," *The Journal of Defense Modeling and Simulation*, vol. 13, no. 3, pp. 331–342, 2016.
[8] C. L. Krishna and R. R. Murphy, "A review on cybersecurity vulnerabilities for unmanned aerial vehicles," in *2017 IEEE International Symposium on Safety, Security and Rescue Robotics (SSRR)*, Shanghai, China, 2017: IEEE, pp. 194–199.
[9] A. Shafique, A. Mehmood, and M. Elhadef, "Survey of security protocols and vulnerabilities in unmanned aerial vehicles," *IEEE Access*, vol. 9, pp. 46927–46948, 2021.

[10] G. K. Pandey, D. S. Gurjar, H. H. Nguyen, and S. Yadav, "Security threats and mitigation techniques in UAV communications: A comprehensive survey," *IEEE Access*, vol. 10, pp. 112858–112897, 2022.
[11] J. Wang, Z. Jiao, J. Chen, X. Hou, T. Yang, and D. Lan, "Blockchain-Aided secure access control for UAV computing networks," *IEEE Transactions on Network Science and Engineering*, vol. 11, no. 6, pp. 5267–5279, 2023.
[12] R. Kumar, A. Aljuhani, P. Kumar, A. Kumar, A. Franklin, and A. Jolfaei, "Blockchain-enabled secure communication for unmanned aerial vehicle (UAV) networks," in *Proceedings of the 5th International ACM Mobicom Workshop on Drone Assisted Wireless Communications for 5G and Beyond*, Sydney, NSW, Australia, 2022, pp. 37–42.
[13] R. Gupta, M. M. Patel, S. Tanwar, N. Kumar, and S. Zeadally, "Blockchain-based data dissemination scheme for 5G-enabled softwarized UAV networks," *IEEE Transactions on Green Communications and Networking*, vol. 5, no. 4, pp. 1712–1721, 2021.
[14] R. Karmakar, G. Kaddoum, and O. Akhrif, "A blockchain-based distributed and intelligent clustering-enabled authentication protocol for UAV swarms," *IEEE Transactions on Mobile Computing*, vol. 23, no. 5, pp. 6178–6195, 2023.
[15] D. Alsadie, "Cybersecurity and artificial intelligence in unmanned aerial vehicles: Emerging challenges and advanced countermeasures," *IET Information Security*, vol. 2025, no. 1, p. 2046868, 2025.
[16] V. P. Gandi, N. S. L. Jatla, G. Sadhineni, S. Geddamuri, G. K. Chaitanya, and A. Velmurugan, "A comparative study of AI algorithms for anomaly-based intrusion detection," in *2023 7th International Conference on Computing Methodologies and Communication (ICCMC)*, Erode, India, 2023: IEEE, pp. 530–534.
[17] T. Sowmya and E. M. Anita, "A comprehensive review of AI based intrusion detection system," *Measurement: Sensors*, vol. 28, p. 100827, 2023.
[18] A. Alharbi, I. Petrunin, and D. Panagiotakopoulos, "Assuring safe and efficient operation of UAV using explainable machine learning," *Drones*, vol. 7, no. 5, p. 327, 2023.
[19] C. S. Onunta and U. B. Chaudhry, "Policy and compliance: Regulatory perspective for autonomous aerial and surface operations," in Umair B. Chaudhry, and Hamid Jahankhani (eds.), *Autonomous Revolution: Strategies, Threats and Challenges*, 2025, pp. 179–198. Springer.
[20] C. Lavers, "Design in engineering: An evaluation of civilian and military unmanned aerial vehicle platforms, considering smart sensing with ethical design to embody mitigation against asymmetric hostile actor exploitation," Technical Report, 2021.
[21] M. Adil, H. Song, S. Mastorakis, H. Abulkasim, A. Farouk, and Z. Jin, "UAV-assisted IoT applications, cybersecurity threats, AI-enabled solutions, open challenges with future research directions," *IEEE Transactions on Intelligent Vehicles*, vol. 9, no. 4, pp. 4583–4605, 2023.
[22] S. K. Shandilya, F. Ortiz-Rodriguez, S. Shandilya, and G. Romero, *AI, Cybersecurity and Data Science for Drone and Unmanned Aerial Vehicles: Real-Life Applications and Case Studies*: CRC Press, 2025.

[23] M. Stampa, A. Sutorma, U. Jahn, J. Thiem, C. Wolff, and C. Röhrig, "Maturity levels of public safety applications using unmanned aerial systems: A review," *Journal of Intelligent & Robotic Systems*, vol. 103, no. 1, p. 16, 2021.

[24] E. Torba and H. Jahankhani, "Securing systems from aerial threats: Cybersecurity in the drone era," in *Autonomous Revolution: Strategies, Threats and Challenges*: Springer, 2025, pp. 261–306. https://doi.org/10.1007/978-3-031-96782-5_9

Chapter 2

Post-quantum cryptography and security models for unmanned aerial vehicles communications

Simar Singh Rayat, Chandradeep Bhatt, Vikrant Sharma, and Susheela Dahiya

2.1 INTRODUCTION

Unmanned Aerial Vehicles (UAVs) have become a part of modern cyber-physical infrastructures. Initially limited to reconnaissance and remote piloted military missions, UAVs are now working as autonomous agents in the fields of precision agriculture, environmental controlling, intelligent transportation, disaster response, smart cities and large-scale logistics. The change of UAV platforms is driven by the improvement of embedded computing, wireless networking, artificial intelligence and distributed control. Consequently, modern UAV systems are no longer stand-alone devices but rather represent nodes in extremely dynamic, heterogeneous and in many cases mission-critical communication networks. The most fundamental element in the operation of UAVs is the communication subsystem. Command-and-control (C2) links are used for control of flight stability and mission execution, telemetry links provide real-time situational awareness, payload data streaming can be used to illicit high-value information, and inter-UAV links are used to provide cooperative behavior in a swarm configuration.

Compromise of these channels can lead to mission failure and loss of assets or, in the case of safety-critical settings, suffer human casualties. Therefore, confidentiality, integrity, authenticity and availability of UAV communications are fundamental and nonoptional [1]. Historically, the security of the communications of UAVs has been based on well-understood cryptographic primitives, such as the RSA key exchange or elliptic curve digital signatures. These mechanisms have been widely used because of their mathematical beauty, mature tools and resistance against conventional enemies. The underlying cryptographic assumptions on which these schemes are based are based on the intricacy of putting numbers like integer factorization and discrete logarithms-all of which fall apart in the presence of sufficiently powerful quantum computers.

The upcoming of quantum computing takes a discontinuity of threat reality. Shor's algorithm has shown that there exist mathematical problems on the foundation of RSA and elliptic curve cryptography that a fault-tolerant

DOI: 10.1201/9781003731580-2

quantum computer has been able to solve in polynomial time. While large-scale quantum computers that can perform such attacks are not currently available to the public, their realization is fairly expected to happen during the operational lifetime of many deployed UAV systems. This mismatch over time between the system's lifespan and the validity of the cryptography determines a severe security dilemma. Accordingly, this chapter claims that new UAV systems will need to be designed with quantum-resilient security models [2]. Retrofitting security after cryptographic malady is neither operationally nor economically viable especially for UAV fleets deployed in unreachable or unfriendly environments. Moreover, the peculiarities of the UAV platforms, of which energy budgets, latency requirements for real-time communication, intermittent connectivity and high mobility are the most important, require security solutions which are fundamentally different from terrestrial network security solutions.

2.2 UNMANNED AERIAL VEHICLE COMMUNICATION SYSTEM EVOLUTION

The communication paradigm used by UAV systems has evolved with the development of autonomy and networking. Early UAV platforms were based on point-of-view radio frequency connections between an aerial vehicle and control station on the ground. By the time security was added, it was usually a matter of proprietary encodings, or symmetric encryption algorithms, with the main threat model being signal jamming or signal interception by foes without sophisticated technology. As deployments with the UAVs grew more complex and network centric, the communication architectures became more and more layered.

The modern UAV systems routinely combine the following:

1. Bottom-up requirements and number of required navigation sensors radar,
2. Bidirectional C2 links (navigation and mission control),
3. Image and video channels (multichannel, high throughput sensors, etc.),
4. Inter-rater reliability between such UAVs, which measures the communication links that enable the cooperative sensing and swarm intelligence,
5. Access to on-premining computing systems or databases through a backhaul connection.

This system has transformed the concept of UAVs to mobile nodes in distributed networks instead of isolated airborne platforms. Consequently, UAV security concerns are increasingly merging with security issues in large-scale cyber-physical systems with strict real-time constraints and adversarial wireless environments. Standardized networking protocols and commercial off-the-shelf components have increased the surface area of attack. While standardization is a way to promote interoperability and scalability, it also introduces well-documented vulnerabilities to UAV systems [3]. Classical cryptographic mechanisms have hence been added to UAV communication stacks to realize authentication and encryption, as well as key management. However, these mechanisms provide an implicit assumption of a threat model that is bounded by classical computation. The increasing utilization of swarms of UAVs only adds up to these difficulties. In swarm configurations, dozens or hundreds of UAVs cooperate with each other in an action based on decentralized communication protocols. Trust relationships develop from binary to collective tied structures, and a single compromised node has the ability to spread false information or malicious commands throughout the swarm. Classical cryptography cannot provide much protection against such a cascade of failures, even before taking into account quantum adversaries.

2.3 FORCES THAT FORCE QUANTUM CONCEIVE COMPUTING PARADIGM SHIFT

Quantum computing brings about a fundamentally different computational paradigm that is based on superposition, entanglement and quantum interference. Unlike classical bits that represent two states, either off or on, quantum bits (qubits) can represent and process many states at the same time. This ability makes possible certain classes of problems to be exponentially faster than is possible on classical machines [4]. From a cryptographic point of view, most important is the impact of quantum computing on public-key cryptography. Shor's algorithm shows that integer factorization and discrete logarithms, which are the basis for the RSA and elliptic curves (two widely accepted public-key algorithms), can be solved in polynomial time Q. Consequently, any system that makes use of these primitives in order for long-term security becomes insecure once the quantum adversaries appear. For UAV systems, the threat is not confined to the real-time of multivariate attacks. An especially insidious risk is store now, decrypt and send information later on. Adversaries are likely to intercept and archive encrypted UAV communications today with the expectation of being able to decrypt the communications at a later date when quantum capabilities become mature. For those missions with sensitive geographical information, surveillance intelligence or proprietary industrial data, retrospective decryption has as much potential to do harm as does real-time compromise. Accordingly, it is not enough to argue that quantum threats can be reacted to. By the time quantum adversaries are operationally viable, the damage might be done. UAV systems, especially those that are required for defense and sensitive infrastructure roles, must have the assumption of the existence of quantum-capable adversaries within its operational preachment [5].

2.4 QUANTUM-ERA THREAT MODEL FOR COMMUNICATIONS FROM UAV

A strict model of threats is necessary for proper design of security architectures. In the quantum world, however, the enemy is not the one that is limited in computational ability but the one with a combination of strategic patience and having access to sophisticated signal intelligence capabilities.

This chapter uses a powerful model of the quantum adversaries with the following assumptions:

1. The opponent has or will have access to large-scale quantum computing resources that are capable of performing cryptanalytic attacks against classical public-key schemes.
2. The adversary has wide-ranging passive monitoring capabilities, which allow them to intercept and store UAV communication traffic for an extended period of time.
3. Direct physical access to UAV hardware is restricted but possible for isolated situations, that is, crash recovery/capture.
4. The adversary may use nodes that have been compromised by UAVs to execute insider attacks in swarm networks.

Under this model, the traditional concept of confidentiality, integrity and authenticity is blurred. Once cryptographic identities are broken, authentication mechanisms fail, making it possible to command and spoof and orchestrate denial of service attacks. In the swarm cases, the consequences are particularly severe. A quantum-enabled adversary that compromises the cryptographic root of trust is able to impersonate the swarm leaders from legitimate sources, give malicious coordination commands and destabilize collective behavior [6]. The resulting system failure is not only a communication breakdown but an implosion of the ability to make autonomous decisions, as shown in Figure 2.1.

2.5 MOTIVATION FOR POST-QUANTUM BEFORE CRYPTOGRAPHY FOR UAV SYSTEMS

Post-quantum cryptography (PQC) aims at finding cryptographic primitives that are quantum resistant. Unlike quantum key distribution, which is implemented using specialized hardware and requires certain assumptions to be made about its physical properties, PQC is software based and can, therefore, be used with existing communications infrastructures. Among the various PQC families, the field of lattice-based cryptography will be the most attractive candidate for integration into the embedded and constrained environments. Schemes like CRYSTALS-XM, CRYSTALS-Kyber and CRYSTALS-Dilithium provide good security guarantees, reasonable

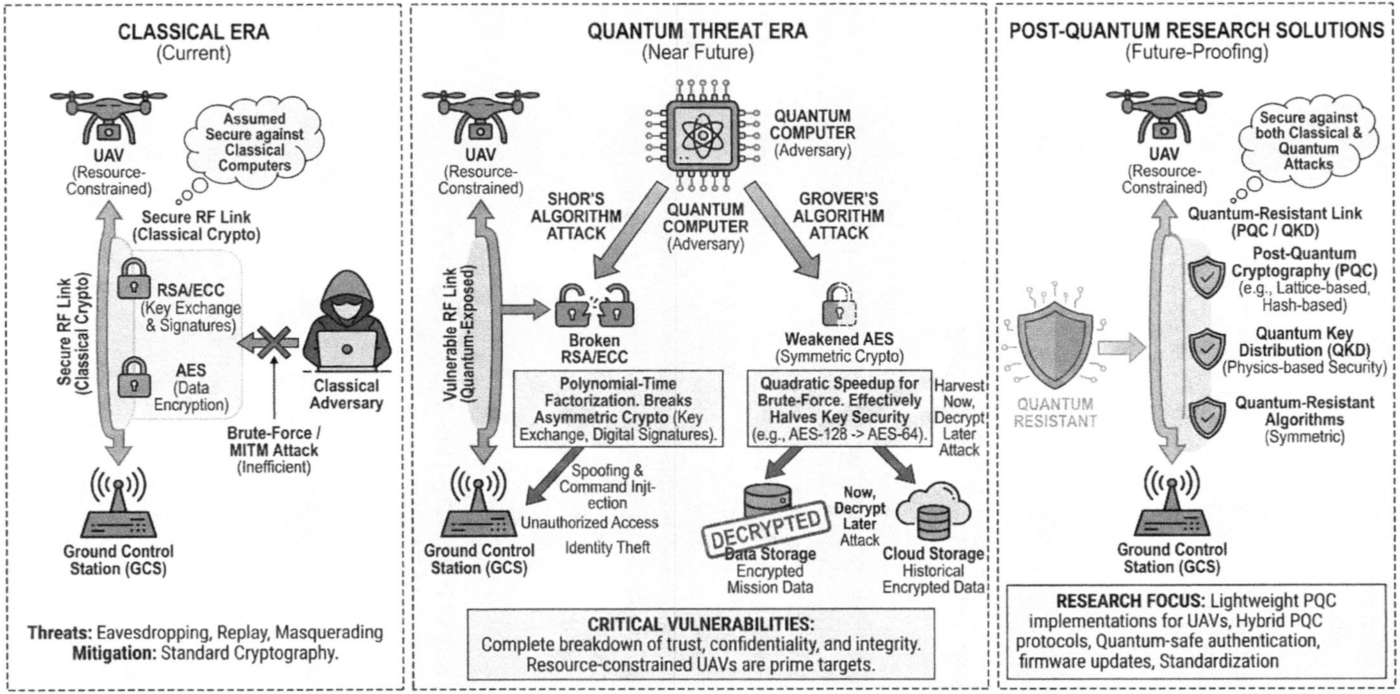

Figure 2.1 Hybrid quantum threat model architecture for UAV communication.

efficiency and flexibility over different classes of platforms [7]. However, the mere substitution of classical algorithms with their PQC counterparts is not a sufficient solution to the issue of the UAV security challenge. PQC, in this case, primitives may consist of larger key sizes and increased computation overheads, which may be incompatible with UAV constraints. Besides, UAV missions are always dynamic, and the levels of threats, connectivity and energy availability change during the missions. Therefore, the motivation behind this chapter is to shift from static application of cryptography to mission-adaptive approach toward system-level security models. Within such models, PQC does not exist as an isolated part of the model, but as an integrated part of a cross-layer architecture that provides the balancing process of security, performance and operational resilience.

2.6 SCOPE AND CONTRIBUTIONS OF THIS CHAPTER

This chapter makes a major advance in the state of the art in three ways. First, it embeds post-quantum cryptography into the operational and architectural peculiarities of UAVs and deviates from the general discussion on network security. Second, it presents a mission-adaptive PQC security architecture that dynamically matches the strength of cryptography with the context of the mission and the threat. Third, it offers a swarm-aware trust and key management model that aims at containing compromise and maintaining the collective autonomy in quantum threat environments. This chapter has created the underpinning motivation and threat landscape, it also conducts in depth a deep dive to post-quantum cryptographic foundations, algorithmic tradeoffs and their adequacy for use in UAV platforms, laying the technical ground on which the novel security architecture can be built upon.

2.7 FOUNDATIONS OF THE POST-QUANTUM CRYPTOGRAPHY

Post-quantum cryptography (PQC) includes cryptography that is used by developing cryptography elements that are resistant to attacks launched using both classical and quantum computation capabilities. Unlike quantum key distribution, which relies on specializations of the physical layer and hardware, PQC is software centric and, as a result, is compatible with existing networking stacks and embedded platforms [8]. This property makes PQC especially appealing for functionality on TLOC (Unmanned-Vehicular-Aircraft systems [UAVs]), where payload mass, power consumption and environmental hardness are factors that do not allow the general use of quantum hardware. The conceptual basis for PQC is computational problems which are believed to resist effective quantum algorithm. Grover's algorithm gives a speedup on the order of the square root for the brute force attack—for symmetric cryptography—but the effect can be countered by relatively small increases in key length. Whereas, Shor's algorithm entirely compromises the difficult aspects of the classical public-key schemes. Consequently, PQC is focused on alternative mathematical structures for which the known best attacks on the mathematical models are computationally infeasible. From a systems perspective, PQC needs to be judged not only from a cryptographic viewpoint but also from an implementation standpoint. UAVs are a constrained execution environment in which algorithmic elegance phage has to operate in tandem with real-time determinism, limited memory and energy efficiency [9]. The rest of this part consists of surveying the main PQC families and assesses their relevance to UAVs communication systems.

2.8 TAXONOMY LANDSCAPES

2.8.1 Lattice-based cryptography

These problems have high worst-case to average-case decreases and hence give great confidence in security assumptions [10]. For UAV systems, cryptographic schemes based on lattices have a number of practical advantages. Applications include the key generation, encapsulation/decapsulation and data encapsulation. Although the key generation and encapsulation operations are efficient in general purpose processors, the operations used are amenable to optimization through vectorization and hardware acceleration. Lattice-based schemes also facilitate key establishment schemes such as digital signatures, which facilitate the use of a unified cryptographic stack. Standardization efforts have taken up such schemes as CRYSTALS-Kyber for key encapsulation and CRYSTALS-Dilithium for authentication. The combination of security and performance in a wide range of parameter sets has been designed into these schemes so that system designers can choose combinations that are suitable for the mission risk and the platform capability. However, lattice-based schemes tend to have larger public keys and ciphers than their classical counterparts, which create an overhead in bandwidth-constrained UAV links, especially those that operate in out-of-line-of-sight. The remainder of this chapter discusses the mitigations of these costs that can be made with architectural design and protocol optimization.

2.8.2 Hash-based cryptography

Hash-Based Cryptography is based on the preimage and collision resistance of cryptographic hash functions. Unlike number theoretic schemes, hash-based signatures are not based on algebraic schemes that can be easily destroyed by quantum algorithms such as Shor's algorithm. As a result, they provide conservative security guarantees. From the implementation point of view, hash-based schemes are attractive because of their conceptual simplicity and very low cryptographic assumptions. They are especially applicable to scenarios that require one-time or infrequent authentication [11]. In UAV situations, hash-based signatures are best suited in scenarios where a secure boot procedure, firmware updates and broadcast authentication tasks are required, where message frequency is low and predictable. The main drawback of using hash-based signatures is state management and signature size. Stateful variants require care to be taken against reuse of keys, a problem for intermittently connected UAV networks, while stateless variants reduce this problem and tend to result in increased signatures, which may affect communication with low-latency requirements.

2.8.3 Code-based cryptography

Code-based cryptography is based on the fact that it is difficult to decode random linear codes, and this has been studied for several decades. Its lasting security due to lack of a successful cryptanalytic breakthrough makes it conservative in its credibility. Nevertheless, the huge public-key sizes inherent in code-based schemes create huge problems for UAVs deployments. In practice, the transmission of large keys over radio links with low bandwidth may have unacceptable delays and also more susceptibility to jamming and interception. While the code-based schemes retain theoretical interest, they are only applicable to the UAV systems when bandwidth is available in abundance and long-term security requirements take precedence over operational constraints.

2.8.4 Shadow cryptography (multivariate polynomial cryptography)

Multivariate cryptography is based on the difficulty of solving systems of multivariate polynomial equations of finite fields. These schemes allow for the signature generation and verification to be carried out quickly, an attractive property for systems that are required to be real-time ones. However, faith in their long-term security has been shaken by repeated cryptanalytic attacks on proposed constructions. For UAV systems, where security failures can have catastrophic consequences, conservative choices of cryptographic standards are preferred. Accordingly, multivariate schemes are by and large deemed as experimental and are not suitable for the primary security functions in mission-critical UAV deployment.

2.8.5 Comparative assessment

Among the considerations of the types of PQC families, lattice-based cryptography stands as the most balanced option for the communication systems on UAVs. It provides a good trade-off in terms of security confidence,

Table 2.1 Comparative analysis post-quantum cryptographic (PQC) schemes for UAV communication

PQC category	*Public key impact*	*Real-time latency sensitivity*
Lattice based	Medium	Suitable
Hash based	Low	Suitable
Code based	High	Conditional suitable
Multivariate	Low	Suitable

computation efficiency and flexibility of implementation. Hash-based cryptography complements the lattice-based ones by covering specific authentication requirements, and other families are either impractical or not mature enough [12]. Accordingly, this chapter considers a lattice-centric PQC model with hash-based solutions where appropriate as the basis for the cryptographic security of UAVs' security architectures, as shown in Table 2.1.

2.9 CRYPTOGRAPHIC PRIMITIVES AND COMMUNICATION FUNCTIONS OF UAV

Figure 2.2 illustrates the UAV communication security; it can be broken down into a number of basic functions: key establishment, authentication, data confidentiality and integrity assurance. PQC primitives have to support these functionalities under stringent constraints using strict operation characteristics. Key establishment is so important. UAVs will often be in dynamic topologies where nodes will join and leave networks as the mission changes. Efficient, low-latency mechanisms for key encapsulation are key for sustainable key encapsulation to keep the channels secure minus efficiently disrupting control loop. Lattice-based key encapsulation mechanisms (KEMs) offer this capability allowing fast session establishment even when the environment is intermittently connected. Authentication lies at the basis of trust relationships among UAVs, ground control stations and also auxiliary infrastructure. Digital signatures have to be verifiable in tight timing budgets, especially in the case of command messages with direct flight behavior effects. Signature schemes like Dilithium provide verification speeds that are compatible with UAV real-teams speed of control on-amount appropriately parameterized [13]. Data confidentiality and integrity are usually provided, thanks to symmetric cryptography after session keys are set up. While symmetric algorithms are not fundamentally broken by quantum computing, key lengths will need to be increased to make up the difference due to the operation of Grover's algorithm. In practice, this adjustment is relatively simple and has negligible overhead compared to public-key operations.

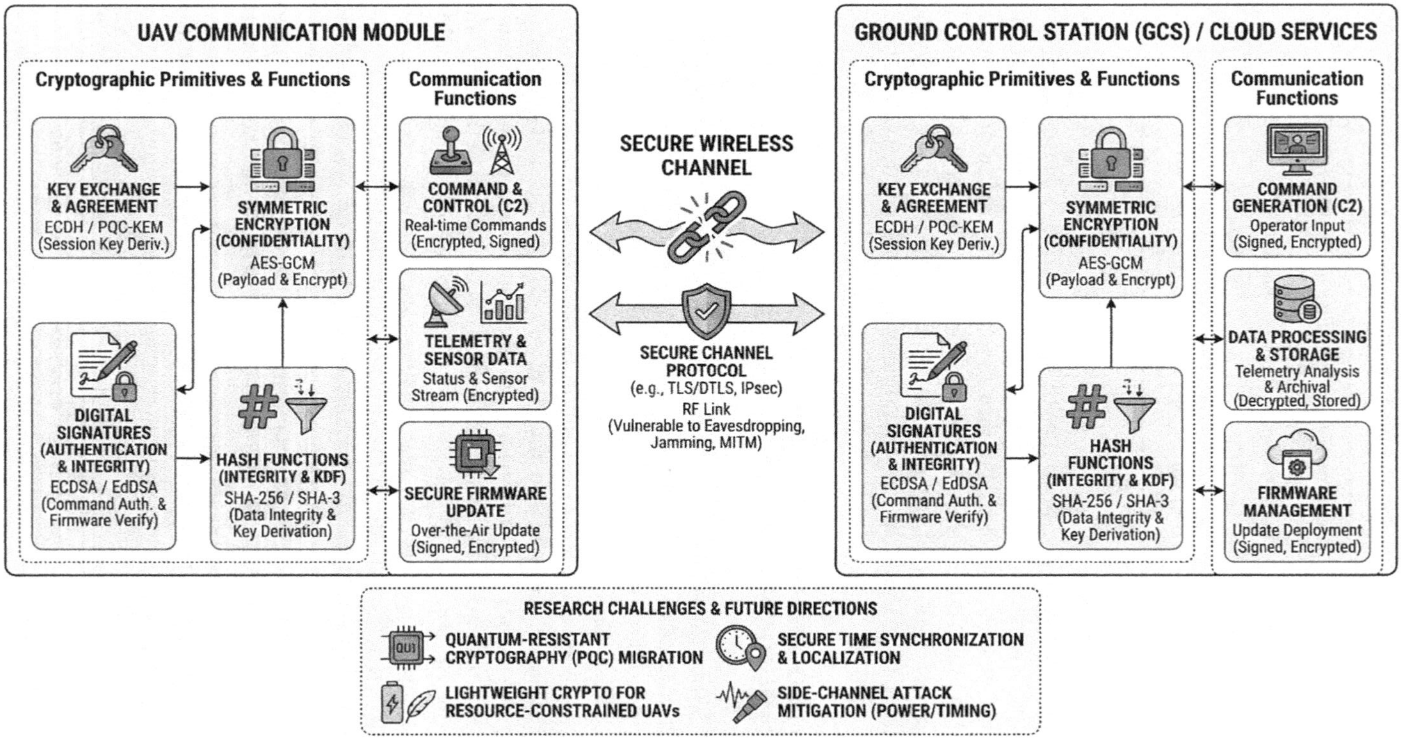

Figure 2.2 Module interaction via communication function and associated cryptographic primitive function overlay in UAVs communication security.

2.10 PLATFORMS FOR UAV AND IMPLICATIONS OF CRYPTOGRAPHY

2.10.1 Computational constraints

UAV onboard processors vary from microcontrollers in small consumer drones to multinuclei system-on-chips in high-end autonomous vehicles. Cryptographic algorithms have to be predictable over this range. Lattice-based PQC schemes can be optimized to not only run efficiently on low-power microcontrollers but on processing devices with higher processing power as well, but attention needs to be paid to ensure that there is no deviation from the expected timing variable nor an excessive use of memory.

2.10.2 Energy constraints

Power consumption is a prevailing design constraint for UAVs. Propulsion is the most significant power consumption, but communication and computation overheads have a direct impact on flight endurance. Cryptographic operations are consequently required to be amortized over mission duration with no place for them during phases of flight where energy margins are tight. This observation provides incentive for session caching, precomputation and mission-free aforementioned cryptography scheduling.

2.10.3 Real time and latency

UAV control loops operate on the millisecond time scale. Cryptographic operations that add indeterminate amount of latency can lead to unstable flight control and subcarrier performance. PQC primitive therefore needs to be integrated in a way that maintains real-time guarantees. Empirical studies suggest that optimized lattice-based key exchange can be performed with acceptable difference in latencies on embedded processors, especially if hybrid handshakes and session resumption techniques are used.

2.11 MAPPING PQC TO THE UAV LAYER OF COMMUNICATION

At the link and network layers, authentication using PQC can make sure that the nodes in the communication are legitimate. At the transport layer, the establishment of secure sessions and protection of data streams against interception and manipulation are secured. At the application layer, the cryptographic authorization provides mission policies and the access control. Crucially, it is not possible to treat these layers separately. Decisions made on one of the layers, for example, the rekeying frequency, have an influence on performance and security on others. This interdependence is the motivation for the cross-layer security architecture presented in the next part of the chapter.

2.12 DRAWBACKS TO ALGORITHM-CENTRIC SECURITY SOLUTIONS

A recurring theme in the study of security for UAVs is a tendency toward the consideration of the cryptographic algorithms in isolation. While the choice of algorithms is important, it does not address such systemic vulnerabilities that arise in protocol designs, in managing trust and in the available operational context. For example, the implementation of a quantum-resistant signature scheme does not make an attacker's life any easier if poor key management or insecure firmware implementation that allows attack to occur [14]. Similarly, strong cryptography cannot overcome poor swarm governance models that allow compromised nodes to have influence on the collective behavior. Accordingly, this chapter rejects exclusively algorithm-centric approaches in favor of system-oriented security models. PQC is seen as an enabling technology used as part of a larger architectural construct that deals with trust, resilience and adaptability.

2.13 SYNTHESIS AND TRANSITION

It has laid down the crypt-secured base required by quantum-resilient fractional private UAV communication systems. It has produced a survey of PQC families, conducted a rationale for (selecting and) lattice-based mechanisms as well as a mapping of cryptographic primitives to UAV constraint and communication functions. The results of the analyses show that PQC is technically feasible in the framework of UAV environments provided sustained by careful architectural design. However, feasibility is not the only requirement lit with security. The next part of this chapter introduces the novel mission-adaptive, cross-layer PQC security architecture that forms the main contribution of this work.

2.14 RATIONALE FOR SYSTEM-LEVEL SECURITY ARCHITECTURE

The previous parts have laid two premises at the heart. First, quantum computing completely undermines security assumptions for classical public needs that form the foundation of most current UAV communications systems. Second, although post-quantum crypto spores are available and technically deployable on the UAV platforms, algorithm substitution is not enough to address the broader security issues for autonomous aerial networks. UAV systems are not fixed endpoints but are mobile, mission aware, cyber-physical agents operating in an adversarial environment with dynamic topology, utilities with dynamic states of energy balance and severe time-of-use constraints [15]. Traditional security architectures spend much time exclusively assuming network conditions relationships are relatively stable, certain credibility relationships and persistent infrastructure. UAV systems contradict all three assumptions. Nodes may periodically disconnect, ground control stations may become inaccessible and trust must be created autonomously in place on many occasions. Under these conditions, cryptographic rigidity becomes a liability, which is an asset. This chapter therefore proposes a novel mission-admissible, cross-layer post-quantum security architecture that is explicitly targeted at UAV communication ecosystems. The proposed architecture integrates the post-quantum cryptography in all layers of the UAV communication stack while permitting cryptographic behavior to evolve dynamically with the assumption of mission context, threat perception and system state. Rather than treating security as a static configuration, it is modelled in a process of continuous control and as a process aligned with the autonomy of the UAV.

2.15 PROPOSED ARCHITECTURE-DESIGN PRINCIPLES

The architecture is governed by five principles of design:

1. Role: Quantum Resilience by Construction. Two possible explanations exist. In the first interpretation, we have two types of resilience. All long-term trust anchors, identity credentials and key establishment mechanisms are post-quantum secure. Classical cryptography is only allowed in hybrid while under controlled migration phases.
2. Mission adaptivity: Cryptographic security, strength, key lifetime and authentication level depend upon the mission phase, threats level and criticality of operations. This approach allows limiting superfluous overhead in case of low-risk operations, ensuring maximum protection in case of delicate phases.
3. Cross-layer coherence: Security mechanisms at the physical, network, transport as well as the application layer are jointly optimized instead of being separately configured. Decisions at one layer say explicitly about behavior at others.
4. Swarm consciousness conservation of trust: The architecture deals with multi-UAV systems as collective entities with a shared security goal. Trust is the subject of hierarchical and local management in order to avoid cascading failures.
5. Crypto-agility and survivability of cryptographic primitives: Parameters and protocols are modular and replaceable, providing fast response to advancements in crypt analysis or compromises in operation. These principles form the unit of differentiation between the suggested architecture and traditional approaches to the security of UAVs and serve as the conceptual basis for the system presented below.

2.16 ARCHITECTURAL OVERVIEW

The architecture being proposed is in the form of a four-level security deck, vertically integrated with the subsystems of UAV autonomy and mission management.

2.16.1 Physical layer and link layer security

At the lowest level, the architecture combines the cryptographic authentication with physical layer level defenses like adaptive frequency hopping, spread spectrum techniques and beamforming. While physical layer security does not substitute the security provided by cryptography, it provides improved security against attacks such as jamming, interception and traffic analysis [16]. Post-quantum authentication tokens are included in the link-layer handshakes, so that authentication is enforced and only authenticated nodes are involved in any form of communications. This avoids rogue UAVs from or malicious relays via the open radio interfaces.

2.16.2 Network and transport layer security

The network and transport layer provides post-quantum key establishment as well as secure session management. Lattice-based KEMs are used to negotiate ephemeral-derived session keys, which are then used in conjunction with symmetric encryption to transfer data. Crucially, there is a context of session establishment. Most of the time (stable connectivity), the lifecycle of the session is longer, thus lower cryptographic overheads. Upon high-risk or high-mobility phases, sessions are reduced and the frequency of rekeying is adjusted upward.

2.16.3 Application and mission security layer

At the application layer, cryptography enforce mission-specific authorization and accessing control. Commands taken for flight behavior, activating payload or control swarm coordination need higher levels of authentication guarantee than generic telemetry, for instance. Mission policies are used declaratively for the definition of cryptographic requirements. For example, entering restricted airspace will automatically initiate enhanced authentication for all the control commands and disable the use of classical cryptographic methods as a backup.

2.16.4 Trust and autonomy integration layer

The uppermost layer combines security and UAV autonomy and decision-making. Trust assessments create impact on planning for initial mission, choosing routes and cooperative behavior. A UAV that identifies anomalous cryptographic behavior from a peer may decrease cooperation, isolate the node or increase the reauthentication requirement. This integration represents a departure away from the historical security model, and cryptography becomes a willing participant in the autonomous control and is no longer a collision membrane.

2.17 MISSION SECURITY ADAPTIVE CRYPTOGRAPHY CONTROL MODEL

One of the ways that the proposed architecture is truly novel is the mission-adaptive cryptographic control model (or architecture). This model assumes that the parameters concerning cryptography are variables that can be controlled by the context of the mission. Mission Phases and Security Posture UAV missions can be broken up into a number of different phases, that is, preflight initialization, takeoff, en-route navigation, task execution and return, each of which has varying threat profiles and performance constraints [17]. High-latency cryptographic operations during initialization are acceptable, which allow a complete authentication and trust establishment. During flight stabilization and avoiding obstacles, there is no room for cryptographic activity to keep the mission up to real-time responsiveness. With sensitive execution of the task, security aspect takes over, and it is worth the extra computational overhead, as shown in Figure 2.3.

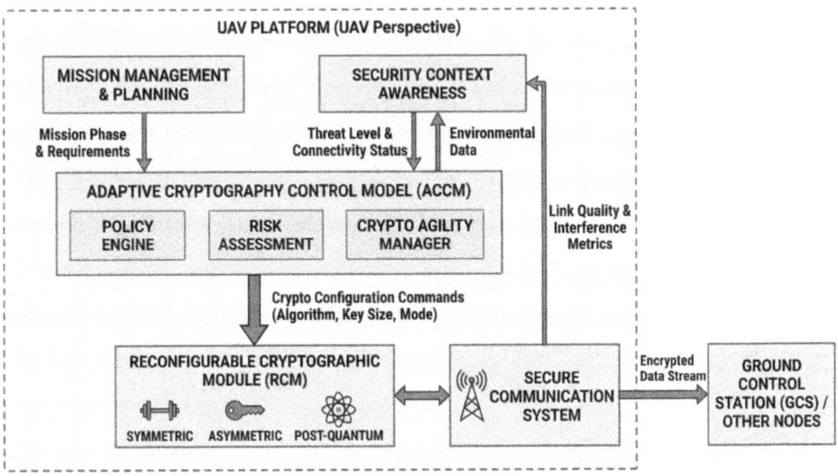

Figure 2.3 UAV communication security adaptive cryptography control model layer orchestration.

2.17.1 Threat adaption of most

Threat perception is based on a range of inputs such as signal anomalies, unexpected topology changes and behavioral deviations of the peer UAVs. These inputs are collated to form a dynamic threat score, which modulates the behavior of cryptography [18]. As the levels of threat increase, the system gradually increases the level of cryptographic survivability (increasing verification frequency to identify signatures and reduce the lifetime of sessions, tighten the check of identities). On the other hand, in nonthreatening conditions, the system conserves resources in terms of energy and bandwidth in terms of relaxed security postures without affecting long-term trust [19].

2.17.2 Cryptographically informed scheduling of energy

The considerations of energy are explicitly taken in account in cryptographic decision-making. In case of energy depletion, the noncritical cryptographic operations are postponed/simplified when the critical, control plane security is kept. This approach will make sure that improvements in security will not compromise flight safety unintentionally.

2.18 SECURE SWARM COMMUNICATION OF TRUST GOVERNANCE

Swarm-based UAV systems pose a completely different security threat than deployments with single vehicles. While trust relations in isolated systems are strictly bilateral, swarm configurations have collective trust relations; failures propagate pretty fast if the containment mechanisms fail.

2.18.1 Hierarchical model of swarm trust

The proposed architecture follows the hierarchical trust model. At the apex, a swarm coordinator, whether a designated leader or a role to be rotated among the participants, is used to establish initial trust anchors which are accomplished using post-quantum credentials. Subgroups within the swarm take over the local trust, thus reducing the need for global coordination [20]. This hierarchy helps to curb the effects of node compromise: whenever a node is compromised by an adversary, the change is made only within the local cluster of nodes, leaving the swarm as a whole relatively resilient.

2.18.2 Group key management

Effective group key management is the key to secure swarm communication. The architecture uses membership-change optimized lattice-based group key establishment protocols. When UAVs enter or leave the swarm, rekeying is done on a local basis, without the high cost of doing it globally. Group keys are used to protect broadcast communication (i.e., formation updates and collective sensing data), and unicast channels are used for sensitive control messages [21].

2.18.3 Identification of compromise, containment of compromise

Swarm members constantly monitor the behavior of their peers for cryptographic aberrant behavior such as invalid signatures or inconsistent identity claims. Suspected nodes are quarantined by consensus-based mechanisms that are decentralized, not based on a single point of authority. This is a spread containment strategy, preventing one compromised UAV from destabilizing the swarm behavior, which would be so within the presence of quantum capable adversaries [22].

2.19 PROTOCOL WORKFLOW ILLUSTRATION

To provide an example of the working flow of the proposed architecture, taking as an example, a standard swarm deployment situation. Prior to starting a mission, each UAV authenticates with the ground control station with postexistential cryptographic credentials and the mission policies are encoded with cryptography requirements. During the takeoff, secure channels are formed with illustrator lattice-based key encapsulation, and initial swarm trust anchors are formed [23]. As the swarm explores to find its way to the mission area, cryptographic activity is kept to a minimum in a bid to maintain energy and responsibility. Upon entry to a high-risk area, when threat indicators are detected, the following escalation is implemented: reduction of session lifetimes, increase in command authentication and group key refresh. If a UAV has anomalous behavior, then the neighboring nodes begin a process of localized reauthentication. Failure of this process leaves only the suspect node isolated without disrupting cooperation of the whole swarm phenomenon. This workflow can be seen to show that cryptography, trust and autonomy do not come and go but are constantly interactive processes instead [24].

2.20 SECURITY ANALYSIS

The proposed architecture provides the strong guarantees for threats due to the quantum era. Confidentiality is provided by post-quantum key establishment and symmetric encryption against quantum brute force attacks. Integrity and authenticity are ensured through adding post-quantum signatures to control and coordination protocols. Availability is supported by cross and within layers of defense along with localized failure containment. Importantly, the architecture deals with not only externals but also insides as the threats from compromised UAVs emanate [25]. The distribution of trust and the limitation of the compromise scope mean that the system will never suffer catastrophic failures even in partial breach conditions.

2.21 DIFFERENCE FROM EXISTING ARCHITECTURE

Existing UAV security frameworks tend to use static cryptographic settings and are focused on each individual communication link. In comparison, the proposed architecture comes with three important differentiators. First, it incorporates systems-level post-quantum security rather than systems that are a protocol add-on. Second, it sees security as a dynamic process that is mission aware. Third, it specifically deals with swarm-level trust in quantum threat. These distinctions are a basis for establishing the architecture of next-generation autonomous aerial systems.

2.21.1 Movement to performance and deployment issues

While the soundness of the architecture is imperative, implementation of the requirements in practice must draw careful attention to performance, energy consumption and migrating from legacy systems. The ability of the proposed architecture is then evaluated and performance trade-offs are modeled, while outlining a feasible migration path toward fully post-quantum UAVs communication infrastructures.

2.22 PERFORMANCE ANALYSIS OF PQC-ENABLED UAV SECURITY ARCHITECTURES

A common theme in post-quantum cryptography (PQC) is the perceived overhead in computation and communication, especially on resource constrained UAV platforms. This section provides a systematic performance evaluation of the integration of PQC in UAV communication systems, in terms of the computational, time delay, bandwidth and energy aspects.

2.22.1 Computational overhead

A modern UAV platform uses heterogeneous computing architectures on board and vary from low-power microcontrol unit to high-power system-on-chip unit with hardware acceleration. According to empirical studies, optimized lattice-based cryptographic operations can be brought within time considerations compatible to UAV control requirements [26]. Key encapsulation and decapsulation are normally finished within few milliseconds on such embedded processors, while signature verification, often the cost-failure, is controllable in terms of speed along with batching and caching strategies. Critically, the mission-adaptive architecture achieves smallest worst case computational overhead, uncoupling the cryptographic in intensity, continuously controlling cyclic manage from its compute. Heavy cryptographic operations are planned to be placed during noncritical periods such as the initialization, loitering or high-altitude cruise, thus maintaining deterministic control behavior during the flight stabilization and obstacle avoidance.

2.22.2 Overhead of communication and bandwidth

Post-quantum public keys and ciphertexts are larger than their classical counterparts, and thus bandwidth utilization on UAV links is an issue. However, this overhead is mostly taken during the session establishment and rekeying; not during continuous data transmission [27]. The architecture overcomes the bandwidth impact by a number of mechanisms. Session keys are reused when the conditions for the threat permit, and this lessens the frequency of handshakes. Group key management in the swarm scenarios localizes the rekeying operations, making sure no unnecessary global updates are performed. Moreover, mission-aware security policies enable the transmittal of noncritical telemetry to make use of established symmetric sessions, averting the repetition of transmitting PQC material. Simulation-based analysis shows that, when amortized over mission duration, PQC-related bandwidth overhead will be a small percentage of total communication load, especially in sensor and payload heavy missions when data streams are the dominant utilization of the link.

2.22.3 Energy consumption impact

For endurance of UAVs, energy efficiency is of the utmost importance. While the energy used for cryptographic functions is high, the energy usage of crypto is comparable when compared to the energy expended for the propulsion, sensing and communication subsystems [28]. Measuring does show a very small percentage due to the cryptographic computation off total energy expenditure if well scheduled. The proposed architecture is well aware of energy when making cryptographic choices. When supplies of energy to the system are limited, critical security functions are given top priority and less critical operations are deferred. This is to ensure that security improvements will not inadvertently compromise the mission completion or flight safety.

2.22.4 Real-time responsiveness

For the response to maintenance in real time is nonborderline in UAV systems. This operation of the mission-adaptive model ensures that the cryptographic operations will not introduce unpredictable latency in control paths. By separating the security relevant functions from time-critical control loops and through the help of precomputation, the architecture maintains the guarantee of real-time functionality even under increased security postures.

2.23 DEPLOYMENT STRATEGY AND MIGRATION STRATEGY

Transitioning existing UAV fleets to some post-quantum security architectures offers both technical and organizational problems. Substitute replacement of old-fashioned encryption is neither practical nor desirable because of the maturity of existing systems and their ubiquity. Therefore, a phased migration strategy among continuity, security and operational feasibility is suggested.

2.23.1 Hybrid cryptographic false deployment end of file (EOF)

In the first stage, UAV systems use hybrid cryptographic techniques that use a mix of classical and post-quanta. Session keys are based on both classical and PQC-based key exchanges in order to ensure that security is retained when at least one part of the exchange is not cracked. This phase enables incremental adoption of PQC while doing the least amount to disburse existing infrastructure and certification processes. It also provides empirical performance data for the parameter tuning and optimization [29].

2.23.2 Phase sustainable development: PQC-dominant operation

Upon achieving confidence in PQC implementations, systems become PQC-dominant operation. Classical cryptography is kept for backward compatibility/low-risk auxiliary functions only. Long-term identities, firmware authentication and control plane security rely entirely on post-quantum primitives. During this phase, mission-adaptative security policies become fully functional, the capability of pushing UAV to cryptographically change dynamic behavior in response to threats and context.

2.23.3 Phase tranquille: crypto totally agile architecture

The final phase will focus on crypto-agility—the capability of swapping cryptographic algorithms and parameters without having to redesign the whole hardware or its system [30]. Cryptographic modules are abstracted behind well-defined interfaces, which allow expected response to future cryptanalysis advances. This phase ensures that UAV security architectures are resilient not only to known quantum threats and the tortuous (e.g., advanced quantum) but also to cryptography and computation.

2.24 CONCLUSION AND FUTURE DIRECTION

This chapter has provided a very in-depth, system-level look at post-quantum cryptography and security models for UAV communications. Moving away from the algorithm (centric nature of the discussion), it proposes a new mission—a novel mission-adaptive, cross-layer security architecture that integrates post-quantum cryptographic primitives, unmanned aerial vehicle (UAV) autonomy, swarm coordination and operations. By being proactive in the holistic approach to quantum-empirical threats, the approach outlined in the suggested methodology lays the foundation for secure, resilient and future proof UAV communication systems. As UAVs become more autonomous networked agents in critical areas of operation, forward looking security architecture focused on critical issues will be a major component of ensuring trust, safety and mission success in the decades to come. Looking into the future, post-quantum security will be one of the pillars of resilient autonomous systems. UAVs are an especially interesting area of research because of their long time periods, operations in hostile environments and the importance to society. Future research effort will most likely be focused on combining cryptography with physical layer security and artificial intelligence (AI)-based threat perception and to utilize distributed ledger technologies for decentralized trust. Advancements in hardware acceleration and low-power computing will continue to lower the cost of quantum resiliency security, thus opening the door for its wider adoption across classes of UAVs.

REFERENCES

[1] M. A. Khan et al., "Future-Proofing Security for UAVs with Post-Quantum Cryptography: A Review," *IEEE Open Journal of the Communications Society*, vol. 5, pp. 6849–6871, 2024.

[2] A. Singh and R. Kumar, "Securing UAV Communications: A Comparative Performance Analysis of Post-Quantum Cryptographic Techniques," *IEEE International Conference on Consumer Electronics (ICCE)*, Las Vegas, NV, 2025.
[3] L. Wang, Z. Liu, and X. Chen, "A Post-Quantum Authentication and Key Agreement Scheme for Drone Swarms," *MDPI Electronics*, vol. 14, no. 17, p. 3364, 2024.
[4] J. Zhang and Y. Li, "A Quantum-Resistant Identity Authentication and Key Agreement Scheme for UAV Networks Based on Kyber Algorithm," *MDPI Drones*, vol. 8, no. 8, p. 359, 2024.
[5] T. Brown et al., "Post-Quantum UAV Communications Encryption Tester (P-QUAVCET): A Framework for Performance Evaluation," *IEEE International Conference on Consumer Electronics (ICCE)*, Las Vegas, NV, 2025.
[6] P. D. Roodsari, S. B. Sarmadi, and H. M. Boorani, "Post-Quantum Authentication and Communication Security for Drone Networks," *IEEE Transactions on Dependable and Secure Computing*, vol. 23, no. 02, pp. 3452–3463, 2026, https://doi.org/10.1109/TDSC.2025.3636937
[7] H. Saini, "GeNp-ODHR: Green Energic and Smart Network Performance Estimation-Based Optimized Deep Hello Routing for Flying Ad Hoc Network," *SAE International Journal of Connected and Automated Vehicles*, vol. 8, no. 3, pp. 319–335, 2025, https://doi.org/10.4271/12-08-03-0022
[8] N. Kumar and A. K. Das, "Lightweight and Quantum-Resistant Authentication for the Internet of Drones (IoD) Using Dilithium Signatures," *IEEE Transactions on Vehicular Technology*, vol. 8, no. 2, pp. 2842–2853, 2024.
[9] S. A. Chaudhry et al., "Anonymous Lattice-Based Authentication Protocol for Vehicular and UAV Communications," *IEEE Internet of Things Magazine*, vol. 48, p. 15, 2024.
[10] L. He, M. Zhao, X. Wang, J. Wang, Z. Wang, and S. Liu, "A Post-Quantum Authentication and Key Agreement Scheme for Drone Swarms," *Electronics*, vol. 14, no. 17, p. 3364, 2025, https://doi.org/10.3390/electronics14173364
[11] Mohammad Wazid, Saksham Mittal, Ashok Kumar Das, S. K. Hafizul Islam, Mohammed J.F. Alenazi, and Athanasios V. Vasilakos, "Designing Secure Blockchain-Based Authentication and Key Management Mechanism for Internet of Drones Applications," *Journal of Systems Architecture*, vol. 160, p. 103365, 2025, https://doi.org/10.1016/j.sysarc.2025.103365
[12] C.-L. Chen, K.-W. Zeng, W.-Y. Li, C.-F. Lee, L.-C. Liu, and Y.-Y. Deng, "Lightweight Post-Quantum Cryptography: Applications and Countermeasures in Internet of Things, Blockchain, and E-Learning," *Engineering Proceedings*, vol. 103, no. 1, p. 14, 2025, https://doi.org/10.3390/engproc2025103014
[13] R. Zhou, H. Guo, F. E. C. Teo, and S. Bakiras, "A Survey on Post-Quantum Cryptography for 5G/6G Communications," *2023 IEEE International Conference on Service Operations and Logistics, and Informatics (SOLI)*, Singapore, 2023, pp. 1–6, https://doi.org/10.1109/SOLI60636.2023.10425346
[14] B. Ojetunde, T. Kurihara, K. Yano, T. Sakano, and H. Yokoyama, "A Practical Implementation of Post-Quantum Cryptography for Secure Wireless Communication," *Network*, vol. 5, no. 2, p. 20, 2025, https://doi.org/10.3390/network5020020

[15] J. Señor, J. Portilla, and G. Mujica, "Analysis of the NTRU Post-Quantum Cryptographic Scheme in Constrained IoT Edge Devices," *IEEE Internet of Things Journal*, vol. 9, no. 19, pp. 18778–18790, 2022, https://doi.org/10.1109/JIOT.2022.3162254

[16] S. Albert, and K. Gunaseelan, "A Lightweight Privacy Preserving Authentication Framework for Secure UAV Communication in IoD Networks," *International Journal of Communication Systems*, vol. 38, no. 13, p. e70187, 2025, https://doi.org/10.1002/dac.70187

[17] H. Park, and Y. Park, "Quantum Secure Authentication and Key Exchange Protocol for UAV-Assisted VANETs," *Mathematics*, vol. 14, no. 5, pp. 820, 2026, https://doi.org/10.3390/math14050820

[18] S. K. Sanadhya et al., "Benchmarking NIST PQC Finalists on UAV Processors," *Proceedings of the International Conference on Security, Privacy, and Applied Cryptography Engineering (SPACE)*, IIT Roorkee, 2023.

[19] Yingpan Kuang, Qiwen Wu, Riqing Chen, and Xiaolong Liu, "Blockchain Based Lightweight Authentication Scheme for Internet of Things Using Lattice Encryption Algorithm," *Computer Standards & Interfaces*, vol. 93, p. 103981, 2025, https://doi.org/10.1016/j.csi.2025.103981

[20] B. Bera, A. K. Das, and B. Sikdar, "Quantum Resistant Lattice-Based Access Control Scheme for UAV-Assisted Internet-of-Drones Applications," *IEEE Transactions on Mobile Computing*, vol. 25, no. 3, pp. 4077–4090, 2026, https://doi.org/10.1109/TMC.2025.3622654

[21] Prithwi Bagchi, Raj Maheshwari, Basudeb Bera, Ashok Kumar Das, Youngho Park, and Pascal Lorenz, "Public Blockchain-Envisioned Security Scheme Using Post Quantum Lattice-Based Aggregate Signature for Internet of Drones Applications." *IEEE Transactions on Vehicular Technology*, vol. 72, no. 8, pp. 10393–10408, 2023, https://doi.org/10.1109/TVT.2023.3260579

[22] C. Rubio Garcia, S. Rommel, J. J. Vegas Olmos, and I. Tafur Monroy, "Enhancing the Security of Software Defined Networks via Quantum Key Distribution and Post-Quantum Cryptography," in R. Mehmood, V. Alves, I. Praça, J. Wikarek, J. Parra-Domínguez, R. Loukanova, I. de Miguel, T. Pinto, R. Nunes, & M. Ricca (Eds.), *Distributed Computing and Artificial Intelligence, Special Sessions I, 20th International Conference*. Lecture Notes in Networks and Systems, LNNS (Vol. 741, pp. 428–437). Springer. https://doi.org/10.1007/978-3-031-38318-2_42

[23] A. Mondal and B. Baidya, "Quantum-Resistant Lattice-Based Cryptography for Secure UAV Communications," *International Conference on Smart Systems and Wireless Communication*, Kalyani, West Bengal, 2024. Springer Nature Singapore.

[24] S. Mishra et al., "A Survey of Post-Quantum Cryptography: Start of a New Race for IoT and UAV Security," *Cryptography*, vol. 7, no. 3, p. 40, 2023.

[25] K. Jain and H. K. Saini, "An Emergency Rescue Framework through Smart IoT LPWAN," *2023 International Conference on Advancement in Computation & Computer Technologies (InCACCT)*, Gharuan, India, 2023, pp. 668–672, https://doi.org/10.1109/InCACCT57535.2023.10141728

[26] National Institute of Standards and Technology (NIST), "FIPS 204: Module-Lattice-Based Digital Signature Standard (CRYSTALS-Dilithium)," 2024.

[27] National Institute of Standards and Technology (NIST), "FIPS 205: Stateless Hash-Based Digital Signature Standard (SPHINCS+)," 2024.
[28] H. K. Saini and K. Jain, "Implementation of DEWMA-Based Hello Packet for AODV to Improve the Performance of FANET with 3D-GMM," *SN Computer Science*, vol. 5, no. 3, p. 321, 2024, https://doi.org/10.1007/s42979-024-02673-z
[29] QNu Labs, "Quantum Secure Drone Communication Platform: Technical Whitepaper on PQC-UAV Integration," 2024.
[30] D. J. Bernstein and T. Lange, *Post-Quantum Cryptography in Embedded Systems*, CRC Press, 2022.

Chapter 3

Ensuring trust through blockchain technology for decentralized cloud-based UAV networks

Sri Nagesh Osri, J. Balaraju, Praveen Kumar Mundlagiri, B.V. Srikanth, and Raja Rao Budaraju

3.1 INTRODUCTION

To support smart mobility and Sustainable Development Goal (SDG)-focused applications, a growing body of research is incorporating the use of decentralized, cloud-based architecture in Unmanned Aerial Vehicle (UAV) ecosystems so that autonomous, scalable, intelligent processes can be facilitated [1]. However, these distributed environments have serious challenges involving trust, data integrity, security and coordination between heterogeneous UAV nodes and service providers. The blockchain technology offers a highly viable solution that has ledgers that cannot be tampered, decentralized identity and transparent consensus mechanisms that are not managed by centralized players. This chapter discusses how blockchain may be used to offer assurance in the cloud-strategy UAV networks that facilitate the sharing of data, task and responsibility assignment in order to provide sustainability and resilience of UAV ecosystems.

3.1.1 Overview of UAV networks and cloud integration

A network of interconnected aerial robots constitutes UAV networks and can be utilized in sensing, communication and decision-making in a broad range of application domains. It is through the integration with cloud computing that these networks can store data in large volumes, do real-time analytics and process with even more capabilities than onboard computer capabilities. Cloud-based UAV systems can facilitate the distributed planning of missions and intelligence and can coordinate a group of UAVs and ground stations with minimal effort [2]. However, latency, security, reliability and trust are also the problems related to using cloud services. The design and dynamics of cloud-integrated UAV networks are the secret to developing scalable, secure and resilient cloud-integrated systems that will support smart mobility and sustainability-oriented applications.

DOI: 10.1201/9781003731580-3

3.1.2 Need for trust and security in decentralized UAV ecosystems

A decentralized UAV ecosystem is anchored on the interplay of numerous autonomous aerial vehicles, ground stations, cloud and third party stakeholders. Trust creation is a paramount need without a centralized control in order to offer credible coordination, exchange of data security and responsible decision-making. UAVs are often forced to operate in dynamic and hostile environments where communication links can be spoofed, data can be altered, unauthorized individuals could be attacked and malicious threats can launch denial-of-service attacks. Moreover, the secret mission data on surveillance, logistics, disaster response and smart mobility must be secured to prevent abuse and invasion of privacy [3]. The security problems are also caused by heterogeneous nature of UAV networks, lack of sufficient onboard computing power and topological changes that occur at high rate. It needs to have mechanisms of trust to ensure that the identities of UAVs are authentic and that their provenance, as well as integrity of task execution, are verified over distributed nodes. Without well-developed trust and security mechanisms, decentralized UAV systems will be vulnerable to derailed missions, safety risks and nonregulatory violations. Therefore, secure, transparent and scalable trust models are the combination need to create a resilient, collaborative and sustainable UAV ecosystem according to the new applications of smart mobility and SDG-oriented applications.

3.1.3 Role of blockchain in enabling secure collaboration

At the center of safe and reliable cooperation in decentralized UAV ecosystems is blockchain technology. Blockchain enables all parties involved in UAVs and services to access a unified and verifiable account of transaction, mission logs and data exchanges through the dispensation of an unchangeable and resistant to manipulation ledger. A decentralized consensus mechanism does not need a single trusted authority, and, consequently, single points of failure are eliminated and the system becomes more resilient [4]. Identity management can be carried out through blockchain to ensure safe authentication of UAVs and operators, and automate mission coordination, access control and incentive mechanisms through smart contracts in a transparent way. Moreover, accountability and auditability that are supported by immutable data provenance are required to meet regulatory compliance and postmission analysis needs. With such abilities, blockchain creates a reliable collaboration framework that facilitates safe information exchange, joint decision-making and joint execution of tasks, which provide reliable and scalable UAV networks with smart mobility and sustainability-related applications. Figure 3.1 presents the role of blockchain in enabling secure blockchain solutions.

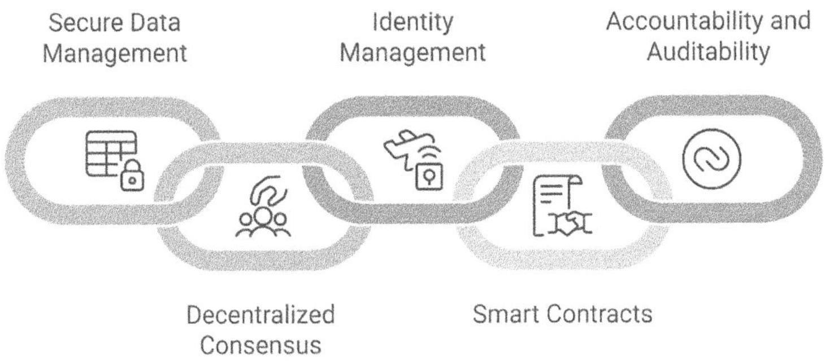

Figure 3.1 Blockchain for secure collaborative solutions.

3.1.4 Chapter objectives and structure

The purpose of this chapter is to analyze how blockchain technology can be used to create a trust, security and transparency in decentralized cloud-based UAV networks. It presents the main security issues, discusses blockchain-based methods of safe cooperation and speaks about the corresponding architectures and applications. The chapter is well-organized in the way that it moves through the introductory ideas to implementation issues, and it ends with the prospective investigation and suggestions about the sustainable UAV ecosystems.

3.2 FOUNDATIONS OF DECENTRALIZED CLOUD-BASED UAV NETWORKS

Decentralized cloud-based UAV networks are a set of distributed aerial platforms that are coupled with cloud and edge computing infrastructure to provide scalable, autonomous and collaborative operations. This is because such networks, unlike centralized architectures, distribute control, data processing and decision-making among UAVs, edge nodes and cloud services and increase fault tolerance and operational flexibility. There are also the ad hoc, mesh and hybrid models of communication that can adapt to dynamic topology. Cloud integration supports intensive data analytics, AI-driven intelligence and long-term storage, whereas edge computing reduces latency on short-latency tasks. When these foundations are integrated, resilient, adaptable and intelligent UAV ecosystems can be underpinned to enable smart mobility and sustainability-based applications.

3.2.1 UAV network architectures and communication models

The UAV network architectures provide the structure in which the aerial platforms, ground stations and the computational facilities are configured to support coordinated operations. Traditional centralized systems are ground control station or cloud server based to conduct mission planning and data processing. These architectures despite being easy to manage are limited to scale, single point of failure and likely to experience increased latency. On the other hand, decentralized and distributed UAV architectures enable UAV communication and collaborative decision-making between the UAVs in a peer-to-peer manner, which makes them more resilient and adaptable to changing environments. Communication networks in UAV networks are UAV-to-UAV (U2U), UAV-to-Ground (U2G) and UAV-to-Infrastructure (U2I). The Flying Ad hoc Network and Ad hoc Network models enable self-organizing and dynamic routing of data by UAVs without any infrastructure, which is why they can be applied in disaster response and remote monitoring [5]. Mesh-based communication can be used to increase reliability by using multihop routing, and hybrid models can be used to introduce cellular networks, satellite links and 5G/6G infrastructure to offer wide-area coverage and high data rates. Cloud and edge computing also affect the UAV communication architecture, in that they allow offloading tasks, distributed intelligence and real-time analytics. Edge nodes (ground stations and mobile edge servers) process latency sensitive information nearer to UAVs, whereas computer intensive tasks and global coordination are done on cloud platforms. The choice of a decentralized cloud-based UAV ecosystem architecture and communication models is an essential factor in the achievement of scalability, reliability, security and the efficient use of resources. Figure 3.2 presents various UAV network frameworks.

3.2.2 Cloud and edge computing in UAV operations

Cloud and edge computing are relevant in terms of improving the efficiency, scalability and intelligence of the current UAV operations. Cloud computing uses virtually unlimited computing capabilities and storage to support large-scale data analytics, long-term data archiving of missions and advanced AI model training to support traffic monitoring, precision agriculture and disaster management applications. By putting the computation-intensive processes in the clouds, UAVs will be able to transcend the onboard resource constraints and increase the range of operations [6]. Edge computing supplements the cloud by moving the calculations closer to UAVs by using the ground stations, base stations or mobile edge servers. This close range greatly minimizes latency and bandwidth usage, which is vital to functions that are time-sensitive like obstacle avoidance, real-time navigation and swarm coordination. The synergistic nature of the cloud and edge computing in decentralized UAV ecosystems allows achieving distributed intelligence, enhanced reliability and adaptive decision-making, which is the foundation of resilient and autonomous UAV networks [7].

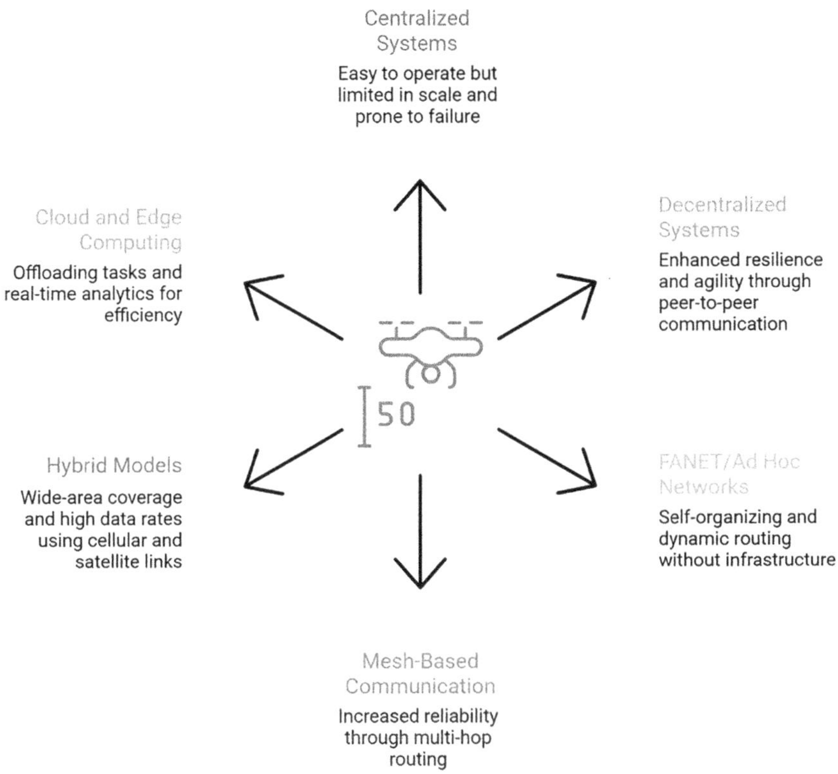

Figure 3.2 Various UAV network architectures.

3.2.3 Decentralization paradigms and limitations of centralized control

The UAV networks paradigm of decentralization changes the control, decision-making and data management that was centralized in one authority to numerous decentralized ones that enhance scalability, resilience and autonomy. Some of the methods include P2P coordination, swarm intelligence and blockchain-based consensus that allow UAVs to coordinate their missions with each other (without a central controller) and exchange information and adjust to evolving conditions. Decentralized architectures reduce single point of failures and increase fault tolerance that is required in large-scale or mission important deployments. However, coordination and decision-making in centralized control schemes are very dependent on ground control stations or cloud servers. Centralized systems are easier to implement and manage, but with latency issues and reduced scalability, they are prone to cyberattacks and operational disruptions in cases of communication failure. Such limitations inhibit real-time responsiveness and power, particularly when it comes to dynamic or adversarial environment. Thus, decentralized paradigms increasingly gain popularity to come up with secure, autonomous and sustainable UAV ecosystems [8].

3.3 BLOCKCHAIN TECHNOLOGY OVERVIEW

Blockchain is a decentralized record keeping technology based on a distributed ledger that allows secure, transparent and tamper-resistant record keeping over a decentralized network. It works based on the cryptography methods and consensus mechanisms that provide the integrity of data without the involvement of a central authority. The blockchain is impervious and verifiable to transactions and thus the technology is highly applicable to creating trust, accountability and safe coordination of decentralized UAV networks and cloud-based collaborative platforms.

3.3.1 Blockchain architecture and consensus mechanisms

The blockchain design is a decentralized network of nodes which together store a secure and immutable registry of transactions. A cryptographic hash is used to connect a block of transactions to the last block, and every block has a set of transactions, a timestamp and a cryptographic hash. Depending on access permission and governance models, the architecture can be divided into public, private and consortium blockchains. Validation and agreement of transactions on distributed nodes without a central authority would not be possible without consensus mechanisms. Some of the common ones are Proof of Work, which is based on computational effort; Proof of Stake (PoS), which is based on stake or ownership and Practical Byzantine Fault Tolerance (PBFT), which is applicable to permissioned networks with low latency and high throughput. Advanced balance mechanisms such as Delegated PoS and hybrid consensus models are balanced with scalability, energy efficiency and security. These architectural and consensus schemes in UAV networks determine the effectiveness, dependability and reliability of the decentralized functionality in the networks, which enable the secure transfer of data, task verification and autonomous coordination between UAVs and cloud-edge resources [9]. Various blockchain architectures and consensus mechanisms are shown in Figure 3.3.

3.3.2 Smart contracts and distributed ledgers

The decentralized systems that are made possible through the blockchain are smart contracts and distributed ledgers. A distributed ledger is a redundant database that is stored in multiple nodes, and the participants are guaranteed of synchronized and irreversible record of transactions. It enables

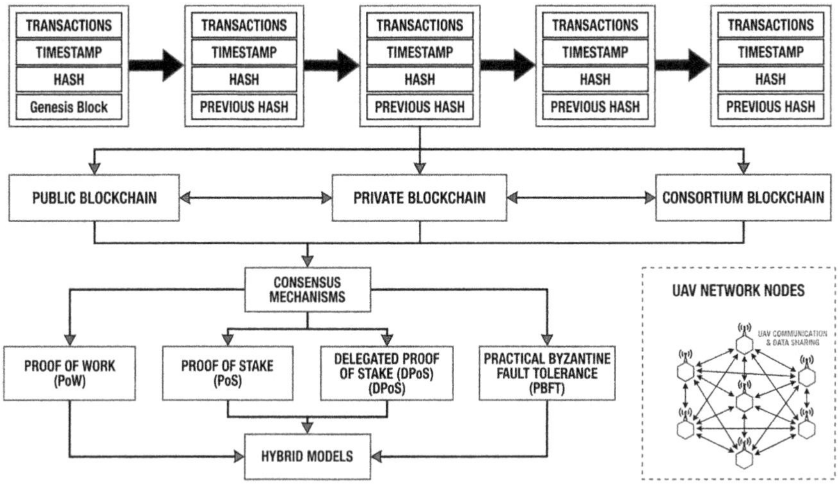

Figure 3.3 Various blockchain architectures and consensus mechanisms.

transparency, traceability and immutability, which are crucial to trust, and accountability to collaborative UAV networks. Smart contracts are computer codes that operate on autopilot and they are stored in the blockchain and execute preset rules and agreements without human intervention. They promote self-scheduling of activities, safe communication of information, access management and auto payment on UAV operation. UAV networks can achieve decentralization of decisions, compliance and reliability and transparency of mission-critical actions with distributed ledgers combined with smart contracts. This will enhance the efficiency of the operations, reduce the centralized aspect of the authority and enhance security within the cloud-based UAV ecosystems [10].

3.3.3 Security features: immutability, transparency and traceability

Blockchain has strong security functions which are required in decentralized UAV networks. Immutability means that once a transaction or data entry is stored on the blockchain, the data cannot be changed or removed and mission logs, sensor data and UAV operational records cannot be changed or modified by unauthorized individuals. Transparency enables all the involved nodes to access and validate transactions, which creates trust and accountability among UAV operators, cloud services and stakeholders. Traceability allows full data provenance, execution of tasks and history of decision-making, which are useful in audits, compliance and analysis of postmission. The combination of these characteristics creates a stable, verifiable and reliable base of the decentralized UAV operations so that cooperative missions are resistant to tampering, accountable and auditable in real-time.

3.3.4 Blockchain platforms relevant to UAV networks

A number of blockchain platforms are specifically well-suited to the use of secure and decentralized UAV networks and each of them has its distinct characteristics based on the specifics of operation needs. Ethernet is a popular open-source blockchain with Turing-complete smart contracts, allowing independent execution of tasks, sharing of data in a secure and safe manner and autonomy in coordinating autonomous UAVs. It is also flexible and has strong developer ecosystem, which is why it is both research and operational UAV deployments, although scalability and transaction costs may be constraining factors. Hyperledger is a permissioned blockchain framework that is aimed at private and consortium networks, has high throughput, low latency and fine-grained access control. It suits enterprise-level UAV ecosystems, where the trusted parties, including logistics providers, regulatory agencies and cloud providers, need to work together safely without revealing sensitive information to the general public [11]. The use of a directed acyclic graph (DAG) named Tangle has allowed Internet of Things Application (IOTA) to use feeless, scalable and lightweight transactions. This renders it suitably applicable to UAV swarms and Internet-of-Things (IoT) integrations, where microtransactions and data sharing, as well as coordination in real-time, are required. The choice of the right platform is based on the size of the UAV network, security needs, maximum latency and mission-critical factors, and this is among decentralization, efficiency and reliability. Figure 3.4 presents various blockchain networks relevant to UAV frameworks.

Table 3.1 presents a summary of the blockchain platform for UAV networks.

3.4 TRUST AND SECURITY CHALLENGES IN UAV NETWORKS

The distributed, dynamic and heterogeneous nature of decentralized UAV networks pose a serious challenge to trust and security aspects. Such risks are the tampering of data, spoofing, unauthorized access and denial-of-service attacks, which can impair the integrity and safety of missions. Without a reliable verification framework, it is hard to build trust among self-directed UAVs, cloud providers and ground controllers. Moreover, due to the onboard constraints, high mobility and dynamic network topologies, real-time authentication, data integrity and secure coordination are especially difficult in UAV ecosystems [12].

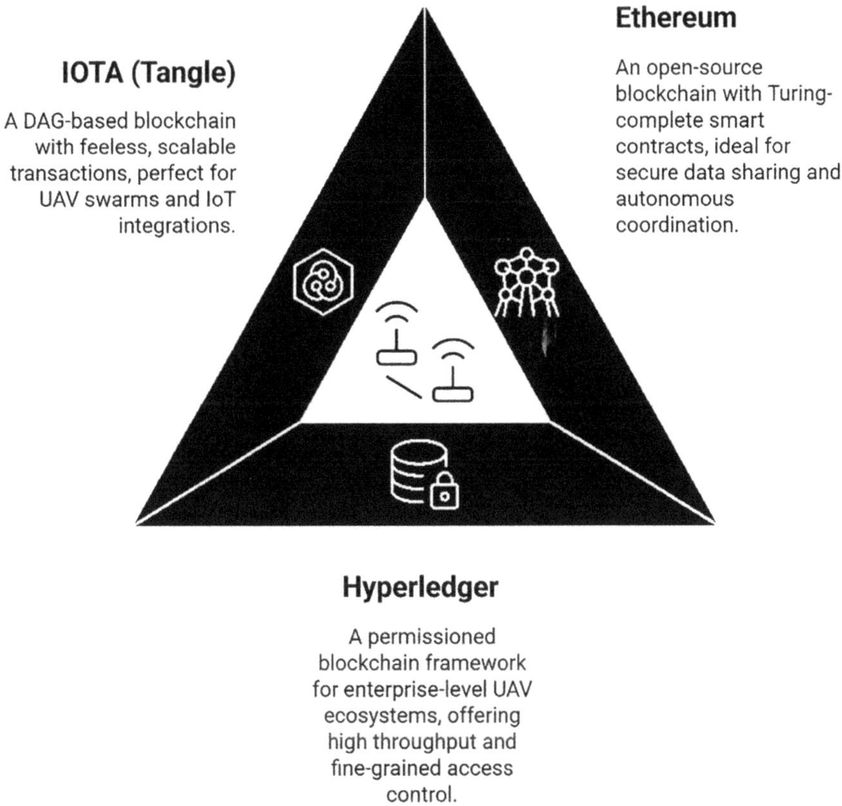

Figure 3.4 Blockchain networks for UAV architectures.

Table 3.1 Blockchain platforms for UAV networks

Platform	*Blockchain type*	*Key features*	*Suitability for UAV networks*	*Limitations*
Ethereum	Public, permissionless blockchain	Turing-complete smart contracts, strong developer ecosystem and high flexibility	Autonomous UAV coordination, secure data sharing and research and pilot deployments	Scalability issues, high transaction costs and latency
Hyperledger (e.g., fabric)	Permissioned, private/ consortium blockchain	High throughput, low latency and fine-grained access control	Enterprise and regulated UAV ecosystems involving trusted parties (logistics, regulators)	Limited decentralization and smaller open ecosystem
IOTA	DAG based (Tangle)	Feeless transactions, lightweight, scalable and IoT oriented	UAV swarms, real-time coordination and IoT and microtransaction-based UAV applications	Less mature ecosystem and evolving security model

3.4.1 Identity authentication and access control

Decentralized UAV networks need a strong authentication of identity to ensure that the only agents involved in mission-critical operations are authorized UAVs, operators and cloud-edge nodes. Authentication schemes check the digital identities with cryptographic keys, certificates or blockchain-based decentralized identifiers (DID) to avoid impersonation and unauthorized access to the system. Access control is the access permissions to data, tasks and network resource usage, which is assigned based on authenticated identities. Role-based, attribute-based and access controls based on smart contracts enable finesse regulation of the operations, so that UAVs and stakeholders are capable of undertaking solely authorized operations. Collectively, both identity authentication and access control can provide a secure base of collaborative UAV ecosystems, such as allowing trusted communication, information exchange and coordinated decision-making in decentralized cloud-based settings. These systems are important to the integrity of their missions, compliance with regulations and protection of sensitive operational data against cyberattacks [13].

3.4.2 Data integrity and privacy concerns

Data integrity and privacy are crucial in decentralized UAV networks, as they are the assurance of security and reliability in the operations. Data integrity ensures that the data collected, communicated or stored, for example, sensor data, flight data and mission data, are accurate, unmodified and verifiable throughout the UAV network. These cyberattacks, communication breakdown or bad actors can ruin this integrity and lead to wrong decision-making or operational risks. The privacy of the collected and shared information (i.e., geolocation, surveillance photos and user data) is also a privacy concern. Unauthorized access or disclosure can be contrary to the laws and can put the stakeholders at a risk. The data can be kept secret and traceable and accountable with such mechanisms as end-to-end encryption, secure multiparty computation and audit trails with the help of blockchain. In order to develop a reliable UAV ecosystems, mission reliability and regulatory compliance and to ensure sensitive operational and environmental data, the issues of data integrity and data privacy have to be taken into consideration [14].

3.4.3 Malicious node detection and trust evaluation

UAV networks that are decentralized can be susceptible to rogue nodes that can interfere with the operations, relay fake data or alter the integrity of the network. Malicious node detection is the process of detecting those UAVs or ground nodes that behave abnormally, for example, inconsistent sensor readings, abnormal communication patterns or unauthorized access attempts. Methods are the detection of anomalies, monitoring of behavior and verifying transactions or the execution of tasks with the help of blockchain. Trust evaluation is a set of measurements used to determine the reliability and reputation of nodes depending on past performance, interaction performance and adherence to network protocols. With the help of assigning trust scores, the system will be able to prioritize communications, data sharing and task assignments to trustworthy nodes and isolate or restrict the impact of untrusted or suspicious participants [15].

3.4.4 Secure communication and mission data sharing

The decentralized UAV networks require secure communication in order to ensure that mission-critical data are not intercepted, manipulated or accessed by an unauthorized third party. The UAVs, ground stations and cloud-edge nodes can be confident in the security of their messages across space and time because of encryption protocols, including end-to-end cryptography and secure tunneling. Authentication schemes are used to authenticate the identity of the senders to prevent man-in-the-middle attacks and spoofing. The mission data sharing allows using blockchain and smart contracts to facilitate transparent, verifiable and tamper-resistant sharing of sensor data, flight data and command data. Decentralized registries store the transactions permanently, and every authorized user can get correct and updated data. Role or attribute access control is used to make sure that sensitive information is distributed among valid parties [16].

The summarization of the various challenges is shown in Table 3.2.

3.5 BLOCKCHAIN-ENABLED TRUST FRAMEWORK FOR UAV NETWORKS

The framework of trust that is created with the help of blockchain creates secure, transparent and decentralized identity verification, data integrity and collaborative decision-making in UAV networks. The framework incorporates distributed ledgers, smart contracts and consensus protocols to provide a system that is tamper-resistant in the recording of transactions, autonomy in the execution of tasks and responsibility in logging missions. This will reduce security threats, facilitate trusted data exchange and create credible exchanges among the UAVs, cloud-edge nodes, and stakeholders in decentralized and mission-critical UAV systems [17].

3.5.1 Architecture of blockchain-based UAV systems

The UAV blockchain systems architecture combines aerial platforms, communication networks and distributed ledger infrastructure to facilitate the

Table 3.2 Trust and security challenges in decentralized UAV networks

Aspect	*Key challenges*	*Security mechanisms*	*Benefits*
Overall trust and security	Data tampering, spoofing, unauthorized access, DoS attacks, high mobility and dynamic topology	Blockchain-based verification, cryptographic protocols and decentralized trust models	Improved mission integrity, resilience and safety
Identity authentication and access control	Impersonation, unauthorized participation and improper resource usage	Cryptographic keys, certificates, decentralized identifiers (DID), role/attribute-based access control and smart contracts	Trusted collaboration, regulatory compliance and controlled operations
Data integrity and privacy	Data manipulation, leakage of sensitive information (location, images and user data)	End-to-end encryption, blockchain audit trails and secure multiparty computation	Reliable data, confidentiality and accountability
Malicious node detection and trust evaluation	Rogue UAVs, false data injection and abnormal behavior	Anomaly detection, behavior monitoring, blockchain verification and trust/reputation scores	Isolation of malicious nodes and reliable coordination
Secure communication and mission data sharing	Interception, spoofing, man-in-the-middle attacks and unauthorized data access	Secure tunneling, encryption, authentication, blockchain-based data sharing and access control policies	Tamper-resistant communication and transparent and secure data exchange

safe and decentralized operations. UAVs are also data producers and consumers, as they create mission data, sensor data and status data. They are relayed in UAV-to-UAV and UAV-to-ground communication connections to edge nodes or ground stations, which communicate with the blockchain network. The blockchain layer is made up of distributed nodes deployed at the edge servers, cloud servers or trusted ground stations, and they maintain the shared ledger, execute smart contracts and support consensus protocols. Smart contracts control authentication of identity, access control, assigning tasks and incentives. Big Data storage, analytics and AI model training

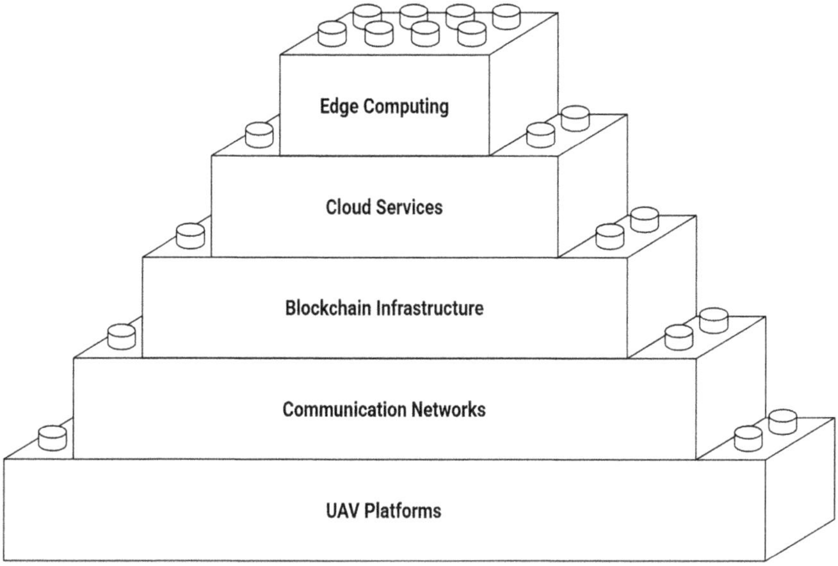

Figure 3.5 Architecture of blockchain-based UAV systems.

can be supported using the cloud layer, whereas low-latency processing of time-sensitive operations is provided by edge computing [18]. The layered architecture can be used to achieve secure data sharing, decentralized trust management and robust coordination, which can be useful in supporting scalable, autonomous and trustworthy UAV ecosystems. Figure 3.5 depicts the frameworks of various blockchain-based UAV systems.

3.5.2 Consensus-driven data validation and node reputation

The validity of information exchanged within decentralized UAV networks is ensured through consensus-based data validation in order to make the information accurate, reliable and tamper resistant. With blockchain consensus mechanisms, several distributed nodes jointly authenticate transactions, sensor data and mission updates and then they are irrevocably stored on the ledger. This joint validation helps to avoid the situation when there is a single node that can inject false or manipulated data to improve data integrity and system reliability. The node reputation mechanisms are used to supplement consensus mechanisms by determining the credibility of UAVs and other entities involved in the process in relation to previous behaviors, data quality and adherence to network protocols. The reputation scores are transparently updated on the blockchain and can be enforced on a smart contract. More reputable nodes are given priority in sharing data and allocation of tasks, whereas suspicious or malicious nodes are limited or separated [19]. Consent-based validation and reputation management combined will form a solid confidence system of safe and robust UAV activities.

3.5.3 Smart contract-based access and task allocation

Smart contracts can be used to provide automated, transparent and secure access control and task allocation in decentralized UAV networks. Smart contracts can be used to determine identities of UAVs, assign roles as well as provide or deny access to data, communication channels and network resources, by encoding predefined rules and policies on the blockchain and not require human intervention. In the assignment of tasks, smart contracts are dynamically allocated missions founded on the UAV capabilities, availability, location and reputation scores. Under the specified conditions, the contracts automatically cause the execution of tasks, track their progress and document the results permanently on the distributed registry. This removes centralized control, minimizes overhead on coordination and prohibits unauthorized activities. Smart contract mechanisms utilize automated enforcement and transparency to improve efficiency, accountability and trust, which allow scalable and reliable collaboration in decentralized cloud-based UAV ecosystems [20].

3.5.4 Case study: blockchain-assisted secure UAV swarm coordination

Take the case of a swarm of UAVs to monitor the destruction after a disaster in the city. Several UAVs work together to survey the impacted regions, gather high-resolution images and transmit real-time information to the emergency responders. A blockchain-based scheme is adopted to make sure that there is secure coordination without centralization. The UAVs have a decentralized digital identity that is registered in a permissioned blockchain. Smart contracts are used to control the formation of swarms, the allocation of tasks and access to common data depending on the capabilities of the UAVs, location and available energy. The sensor information and mission updates are authenticated by the consensus of the edge-based blockchain nodes and are registered permanently, eliminating the possibility of fake data injection and spoofing by malicious UAVs. A reputation system constantly reviews the actions of the UAVs; nodes that provide reliable and credible information receive a better score on trust and are used to perform tasks of high importance. Bad or bad actors UAVs are automatically quarantined by execution of smart contracts. The outcome is a robust, transparent and secure swarm coordination system that ensures operational integrity even in the case of a communication failure or a cyberattack. The case shows the improvement of trust, autonomy and reliability of decentralized UAV swarm operations through blockchain [21].

3.6 INTEGRATION WITH CLOUD AND EDGE INFRASTRUCTURES

The combination of blockchain-based UAV networks with cloud and edge computing networks improves the scalability, responsiveness and computation. Edge nodes are used to enable low-latency blockchain processes, real-time data validation and mission-critical decision-making near UAVs, whereas cloud platforms are used to support large-scale ledger storage, analytics and AI model training. This combination provides a balance between performance and decentralization that allows to conduct safe, effective and robust UAV activities in dynamic and large-scale settings [22].

3.6.1 Cloud-blockchain synergy for data storage and processing

The interaction between cloud computing and blockchain technology is essential in the assisted data storage and processing in decentralized UAVs that are scalable and secure. The data integrity, provenance and access control offered by blockchain are achieved by storing hashes of mission data, transactions and smart contract execution on an immutable distributed registry. This makes it transparent and difficult to tamper with and does not congest the blockchain with huge amount of data. Cloud systems supplement this with elastic storage and high-performance computing to process large-scale UAV data including imagery, sensor stream and AI analytics. Cloud-based off-chain storage allows managing the workloads that demand large amounts of data efficiently, whereas blockchain-based references and permission help to ensure safe access and traceability. This hybrid solution provides a balance among security, scalability and performance, making it possible to share data reliably, and real-time analytics and long-term storage of cloud-integrated, decentralized UAV ecosystems [22].

3.6.2 Edge computing for low-latency transaction validation

The edge computing is critical toward ensuring low-latency transaction validation in blockchain-based UAV networks. Transaction processing and consensus can be done nearer to UAVs by placing blockchain nodes, or lightweight validators, at edge locations, for example, ground stations, base stations or mobile edge servers. This is very close to eliminate delays in communication as compared to cloud-only validation. The Edge-based validation provides real-time authentication, fast data validation and timely execution of smart contracts to perform mission critical functions like collision avoidance, swarm coordination and dynamic task allocation. It also saves bandwidth by filtering and aggregating data prior to transmitting to the cloud. With edge computing and blockchain, the UAV systems can respond faster and provide high reliability and decentralized trust management in dynamic and time-sensitive operational settings [23].

3.6.3 Hybrid cloud-edge blockchain frameworks for UAV networks

Hybrid cloud-edge blockchain systems are a combination of the advantages of decentralized trust, low-latency processing and scalable computation to facilitate advanced UAV network processes. The edge nodes in these structures are the lightweight blockchain clients or validators that perform time-sensitive operations, including transaction verification, identity verification and smart contract execution in close proximity to the UAVs. This provides quick reaction and effective coordination of dynamic environments. The cloud layer supplements edge operations by storing entire blockchain ledgers, running elaborate consensus operations and storing and training AI models on large-scale data. Mission data is sensitive information that is stored on the cloud, and cryptographic hash and access policy are stored on the blockchain to guarantee integrity and traceability. This is a balanced approach based on performance, scalability and security that can be used to create robust, efficient and reliable UAV networks that can support smart mobility and SDG-oriented applications [24].

3.7 PERFORMANCE EVALUATION AND COMPARATIVE ANALYSIS

This section measures blockchain-supported UAV networks based on the following metrics: latency, throughput, scalability, security overhead and energy consumption. The comparison of the approach with the centralized and nonblockchain ones reveals the advantages of trust, resilience and fault tolerance and also reveals the trade-offs concerning computational complexity and system performance [25].

3.7.1 Metrics for security, latency and scalability

The metrics used in assessing blockchain-based UAV networks must be clear metrics on the dimensions of security, latency and scalability. The metrics of security can be authentication success rate, accuracy of data integrity verification, antiattack strength (e.g., spoofing and Sybil attack) and malicious node detection efficiency. End-to-end communication delay, confirmation time of transactions and time required to execute smart contracts are the primary measures of latency in time-sensitive UAV missions. Scalability metrics determine system performance when the number of UAVs, transactions or amount of data are increased, such as throughput, consensus overhead and resource utilization. Collectively, these metrics can give a holistic ground on evaluating the efficacy, strength and feasibility of blockchain-based decentralized UAV networks [26].

3.7.2 Simulation tools and experimental platforms

The performance and practicability of blockchain-facilitated UAV networks require simulation tools and experimental platforms to analyze their performance. UAV simulators like NS-3, OMNeT++ and Gazebo allow to model aerial mobility, communication protocol and network dynamics under controlled conditions. Simulators such as Ethereum testnets, Hyperledger Fabric and IOTA can be used to experiment with consensus mechanisms, smart contracts and workloads of transactions. Experimental platforms combine real or simulated UAV equipment with edge and cloud platforms and allow checking the latency, scalability and security in realistic settings. A combination of simulation and testbed-based experimentation in terms of analysis enables a thorough examination of system behavior and trade-offs and the practicality of decentralized UAV solutions in reality [26].

3.7.3 Comparative study with nonblockchain UAV frameworks

A comparative analysis of blockchain-based and nonblockchain UAV systems identifies the advantages and disadvantages of decentralized trust systems. The conventional UAV systems have centralized control to perform authentication, data management and coordination of tasks, which is easy to design, but creates single points of failure, reduced scalability and susceptibility to cyberattacks. On the contrary, blockchain frameworks promote trust, data integrity, transparency and fault tolerance in a decentralized consensus mechanism and in immutable ledgers. Nonetheless, blockchain integration implies the introduction of new computational and communication load, which may cause a rise in latency and energy waste. Nonblockchain systems might be faster in responding to small scale or mission critical in latency, but they do not have strong trust and accountability processes [27]. This analogy shows that blockchain-based UAV networks are more applicable to large-scale, multistakeholder and security sensitive applications, and the optimization of hybrid designs to balance performance and trust is necessary. Table 3.3 presents a summarization of various blockchain and nonblockchain-based UAV systems.

Table 3.3 Comparison of various blockchain and nonblockchain-based UAV systems

Aspect	*Blockchain-based UAV systems*	*Nonblockchain UAV systems*
Control architecture	Decentralized	Centralized
Trust mechanism	Distributed consensus and immutable ledger	Central authority
Security	High data integrity, transparency and fault tolerance	Vulnerable to single-point failures and cyberattacks
Scalability	Better for large-scale, multistakeholder systems	Limited scalability
Latency	Higher due to consensus and validation overhead	Low, suitable for time-critical missions
Computational load	High (cryptography and consensus)	Low
Energy consumption	Increased energy usage	Energy efficient
Accountability	Strong auditability and traceability	Limited accountability
Suitability	Security-sensitive and collaborative UAV networks	Small-scale, latency-critical operations
Key limitation	Performance overhead	Weak trust and resilience

3.8 EMERGING TRENDS AND FUTURE DIRECTIONS

The development of blockchain-based UAV networks is being influenced by the fast development of artificial intelligence, edge intelligence and next-generation communication systems. Among the trends is the fact that AI-based management of trust is being integrated, and machine learning models are used to dynamically evaluate the behavior of nodes, forecast security threats and modify blockchain parameters, including consensus selection and transaction validation approaches. This improves scalability and responsiveness of large swarms of UAVs. The other prospective path is the implementation of lightweight and energy-efficient blockchain protocols that suit resource-constrained UAVs such as DAG-based ledgers and hybrid consensus mechanisms. The objective of these approaches is to minimize the latency, energy and communication overhead and ensure high security levels. The integration of 6G networks with satellite communications and UAV swarms will also increase the operational distance and stability, allowing real-time coordination of the world with blockchain-based trust systems [28]. Furthermore, cross-chain and interoperable blockchain solutions will likely be used to enable the cooperation of heterogeneous UAV systems, service providers and regulatory authorities. The sustainability point of view is that blockchain-based UAV networks will continue to serve SDG-inspired applications, for example, precision agriculture, disaster response, environmental monitoring and smart logistics. Regulatory compliance, ethical considerations and standardization are also areas that future research has to focus on in order to make deployment safe and accountable. All in all, the future development of hybrid cloud-edge blockchain systems will be essential in the development of secure, scalable and smart UAV ecosystems.

3.9 CONCLUSIONS

Today, in the era of the fast digitalization and growing dependence on autonomous systems, trust, security and transparency have become the primary demands of next-generation UAV ecosystems. The chapter has discussed the central role of the blockchain technology in resolving these problems in decentralized cloud-based UAV networks. With the increase in UAV deployments and multiplicity of stakeholders, the conventional centralized control models are no longer satisfactory to manage the security weaknesses, single-point of failure and the complexity of coordination. The blockchain provides a strong decentralized identity management model, data sharing that is resistant to tampering and a transparent cooperation of UAVs, cloud services and edge infrastructures. With the use of smart contracts, consensus-based validation and reputation-based trust, blockchain-based UAV systems can be used to automate access control, task distribution and mission execution and provide accountability and auditability. The combination of blockchain and cloud and edge computing can further improve the scalability and responsiveness of the system, allowing real-time and data-intensive operations of UAVs. Although blockchain adds to the computational and communication load, a comparative analysis shows that blockchain technology has provided the greatest gains in fault tolerance, resilience and immunity to cyberthreats. In the future, the development of lightweight blockchain protocols, AI-based trust assessment and interoperable architecture will be essential to the feasibility of adoption. Generally, blockchain can be identified as a major facilitator of safe, autonomous and sustainable UAV networks that reinforce clever mobility and SDG-based innovations.

REFERENCES

[1] C. Hong and D. Shi, "A cloud-based control system architecture for multi-UAV," in *Proceedings of the 3rd International Conference on Robotics, Control and Automation*, Chengdu, China, 2018, pp. 25–30.

[2] S. Djeradi, T. Bendouma, N. Lagraa, C. A. Kerrache, A. Lakas, and B. Brik, "Toward the integration of UAVs' services into the cloud," *IEEE Communications Standards Magazine*, vol. 5, no. 4, pp. 25–32, 2022.

[3] R. Sarenche, F. Aghili, T. Yoshizawa, and D. Singelée, "DASLog: Decentralized auditable secure logging for UAV ecosystems," *IEEE Internet of Things Journal*, vol. 10, no. 23, pp. 20264–20284, 2023.

[4] M. Shen, J. Duan, L. Zhu, J. Zhang, X. Du, and M. Guizani, "Blockchain-based incentives for secure and collaborative data sharing in multiple clouds," *IEEE Journal on Selected Areas in Communications*, vol. 38, no. 6, pp. 1229–1241, 2020.
[5] I. Jawhar, N. Mohamed, J. Al-Jaroodi, D. P. Agrawal, and S. Zhang, "Communication and networking of UAV-based systems: Classification and associated architectures," *Journal of Network and Computer Applications*, vol. 84, pp. 93–108, 2017.
[6] P. McEnroe, S. Wang, and M. Liyanage, "A survey on the convergence of edge computing and AI for UAVs: Opportunities and challenges," *IEEE Internet of Things Journal*, vol. 9, no. 17, pp. 15435–15459, 2022.
[7] W. Chen, B. Liu, H. Huang, S. Guo, and Z. Zheng, "When UAV swarm meets edge-cloud computing: The QoS perspective," *IEEE Network*, vol. 33, no. 2, pp. 36–43, 2019.
[8] G. W. Cuillier, "Advantages and disadvantages of centralized versus decentralized information systems and services from a project management perspective," Electronic Thesis, projects, and dissertations, 1487, 2022.
[9] N. Sethi, "Blockchain architecture and consensus mechanisms: Challenges, innovations, and future trends," *Int J Comput Appl ICAIDSC*, vol. 3, pp. 28–35, 2023.
[10] J. Li and M. Kassem, "Applications of distributed ledger technology (DLT) and blockchain-enabled smart contracts in construction," *Automation in construction*, vol. 132, p. 103955, 2021.
[11] H. Saini, "GeNp-ODHR: Green energic and smart network performance estimation-based optimized deep hello routing for flying Ad Hoc network," *SAE International Journal of Connected and Automated Vehicles*, vol. 8, no. 3, pp. 319–335, 2025, https://doi.org/10.4271/12-08-03-0022
[12] Y. Zhi, Z. Fu, X. Sun, and J. Yu, "Security and privacy issues of UAV: A survey," *Mobile Networks and Applications*, vol. 25, no. 1, pp. 95–101, 2020.
[13] M. A. Khan et al., "Security and privacy issues and solutions for UAVs in B5G networks: A review," *IEEE Transactions on Network and Service Management*, vol. 22, no. 1, pp. 892–912, 2024.
[14] N. Kaur, K. Bhardwaj, H. K. Saini, S. Kumari, S. Kumar, and P. Kaur, "Preventing Ethereum blockchain re-entrancy attacks using smart mutex lock sum," in *2023 4th International Conference on Computation, Automation and Knowledge Management (ICCAKM)*, Dubai, United Arab Emirates, 2023, pp. 1–6, https://doi.org/10.1109/ICCAKM58659.2023.10449645
[15] C. A. Kerrache, A. Lakas, N. Lagraa, and E. Barka, "UAV-assisted technique for the detection of malicious and selfish nodes in VANETs," *Vehicular Communications*, vol. 11, pp. 1–11, 2018.
[16] A. Zhang, J. Chen, R. Q. Hu, and Y. Qian, "SeDS: Secure data sharing strategy for D2D communication in LTE-Advanced networks," *IEEE Transactions on Vehicular Technology*, vol. 65, no. 4, pp. 2659–2672, 2015.
[17] X. Jian, P. Leng, Y. Wang, M. Alrashoud, and M. S. Hossain, "Blockchain-empowered trusted networking for unmanned aerial vehicles in the B5G era," *IEEE Network*, vol. 35, no. 1, pp. 72–77, 2021.
[18] Y. Tan, J. Liu, and N. Kato, "Blockchain-based lightweight authentication for resilient UAV communications: Architecture, scheme, and future directions," *IEEE Wireless Communications*, vol. 29, no. 3, pp. 24–31, 2022.

[19] X. Zhan, Z. Liu, J. Yan, D. Lin, and C. C. Loy, "Consensus-driven propagation in massive unlabeled data for face recognition," in *Proceedings of the European Conference on Computer Vision (ECCV)*, 2018, pp. 568–583.

[20] V. Jain and B. Kumar, "Combinatorial auction based multi-task resource allocation in fog environment using blockchain and smart contracts," *Peer-to-Peer Networking and Applications*, vol. 14, no. 5, pp. 3124–3142, 2021.

[21] S. H. Alsamhi et al., "Blockchain-empowered security and energy efficiency of drone swarm consensus for environment exploration," *IEEE Transactions on Green Communications and Networking*, vol. 7, no. 1, pp. 328–338, 2022.

[22] Q. Tang, Z. Fei, J. Zheng, B. Li, L. Guo, and J. Wang, "Secure aerial computing: Convergence of mobile edge computing and blockchain for UAV networks," *IEEE Transactions on Vehicular Technology*, vol. 71, no. 11, pp. 12073–12087, 2022.

[23] M. F. Álvarez, "Edge computing security approaches and their influence on latency reduction in e-commerce payment networks," *Journal of Artificial Intelligence and Machine Learning in Cloud Computing Systems*, vol. 5, no. 11, pp. 1–8, 2021.

[24] E. H. Abualsauod, "A hybrid blockchain method in internet of things for privacy and security in unmanned aerial vehicles network," *Computers and Electrical Engineering*, vol. 99, p. 107847, 2022.

[25] M. Islam, M. Rahman, M. Ariff, H. Ajra, Z. Ismail, and J. Zain, "Blockchain-enabled cybersecurity provision for scalable heterogeneous network: A comprehensive survey," *Computer Modeling in Engineering & Sciences*, vol. 138, no. 1, p. 43, 2024.

[26] A. Yazdinejad, R. M. Parizi, A. Dehghantanha, H. Karimipour, G. Srivastava, and M. Aledhari, "Enabling drones in the internet of things with decentralized blockchain-based security," *IEEE Internet of Things Journal*, vol. 8, no. 8, pp. 6406–6415, 2020.

[27] R. Karmakar, G. Kaddoum, and O. Akhrif, "A blockchain-based distributed and intelligent clustering-enabled authentication protocol for UAV swarms," *IEEE Transactions on Mobile Computing*, vol. 23, no. 5, pp. 6178–6195, 2023.

[28] P. Alvares, L. Silva, and N. Magaia, "Blockchain-based solutions for UAV-assisted connected vehicle networks in smart cities: A review, open issues, and future perspectives," *Telecom*, vol. 2, no. 1, pp. 108–140, 2021.

Chapter 4

Aerial intelligence

Orchestrating cloud and edge computing paradigms in unmanned aerial vehicle infrastructures

Sandeep Gupta, Udit Mamodiya, and M. Shamim Kaiser

4.1 INTRODUCTION

Simultaneously, the spread of these civilian and commercial non-combatant unmanned aerial vehicles (UAVs) is beginning to influence the allocation and consumption of computing resources. That is mainly because the traditional cloud computing architecture that sandwiched generational-advantaged computation capacity and storage capability fails to address the latency-sensitive UAV applications' need for real-time data processing, autonomous decision-making, and response promptness. This type of pull between breadth computational wealth and distribution and operational relevance has led to the attitude of edge computing paradigms for UAVs: a state where, with respect to some sensor and information dispersion within the network, it can benefit from some computation somewhere in the periphery that provides calculations as close as possible to those sensors, thus avoiding delay, unnecessary bandwidth consumption, and improving system resiliency. Satellite information coming from different types of satellites (high-resolution cameras, multispectral imaging, thermal controllers, LiDAR laser systems, and hyperspectral scanners) has developed in the last 20 years, and the industrial applications have grown exponentially. The data processing and analysis become a big issue since terabytes of raw sensor data could have been collected during the flight in commercial inspection of infrastructure with UAVs (Li et al., 2013). The weight of uncompressed 4K video is notoriously high for most types of UAVs, and, in a network where communications are down or inaccessible in remote or disaster-affected settings, it can quickly become the bottleneck in the wireless transport node.

Cloud computing has almost unlimited computational capabilities, making it suitable for complex machine learning model designs and central data processing; local edge computing supports the purpose of ultra-low-latency computation, bandwidth minimization, and seamless access to the network with intermittent or irregular connectivity. On the one hand, the tradeoff between high computational resources and power/thermal constraints, as well as wireless connectivity limits in terms of latency, is a major improvement in autonomous navigation, real-time object detection, and intelligent

DOI: 10.1201/9781003731580-4

decision-making without regard to wireless connectivity. The latest deep learning pipelines can be operated directly at the UAV to perform different visual recognition-swapping operations with numbers not in seconds, but in tens, with clouding. These capabilities are vital for a variety of industries (precision agriculture, infrastructure inspection, emergency response, and environmental monitoring) and can make the difference between mission success and failure. For example, from a computer vision perspective, the real-time identification of human survivors has resulted in quicker response times rather than forwarding a video clip to remote analysts (McEnroe et al., 2022).

Furthermore, with the emergence of mobile edge computing, there is a great opportunity for its exploitation by the provision of a network and reporting computation service capable of expanding the network's IT computational assistance to underserved or unserved regions." UAVs can be used as fly-by base stations and can even become mobile or temporary base stations, which not only collect data for IoT sensors but also provide computational assistance to minimize the processing load on resource-constrained ground stations. For example, an edge intelligence coupled to edge intelligence could gather data from decentralized soil moisture monitoring systems, analyze the collected data, and make irrigation decisions on demand to achieve temporal independence from a remote cloud network (Callegaro & Levorato, 2021).

The recharged, constrained UAVs have a limited amount of battery power for propulsion, sensing, and communication tasks, and they must reveal their remaining battery power to use electricity. Once communication costs associated with offloading tasks to remote servers are factored in, the FTC tradeoff must be well thought out, and it opens a can of worms because there are several variables, such as data size, complexity of computation, network availability, and degree of urgency for mission completion. The UAV operation is also highly mobile, resulting in a temporal shift in the channel's computation load and mission tasking, as well as resource management that can adapt to such dynamism. Unmanned platforms roaming around public spaces gather critical data that must be shielded from processing, broadcasting, and storage, which has made security and privacy more difficult. The operational environment in which UAV-based computing systems must operate in differing application domains that are not similar in terms of their requirements and constraints.

Predefined routes are common in the infrastructure inspection process, and the same RA for computation loads and communication is also intended. In other situations, drones may need to make decisions in non-standardized environments, which leads to a lot of confusion when making decisions under uncertainty. Several other firms are also targeting such generalized areas of online demand for sporadic data and offline processing. Various UAVs are gathering all sorts of nature reports, which may necessitate some complicated data exchange and delivery to process. The challenge of micro-UAV and macro-UAV outdoor applications of tactical dynamic scenes should be addressed, as well as the fact that transferability, availability, and low cost are all important.

4.2 ARCHITECTURAL FRAMEWORKS AND DESIGN PRINCIPLES

The development of UAV-based computing systems should address several aspects, including computation demand, energy conservation, information overhead tracking, and scalability. In the new literature, drones are often seen as part of a hierarchical structure of computing resources, from device (processors) to edge (fog nodes and ground control stations) and cloud layers (remote data centers). Several layers of this network architecture support intelligent offloading of tasks (Faraci et al., 2020) that have a dot-like or lattice-created structure that corresponds to various aspects such as delay, available bandwidth, power, and computational complexity of the task-processing. Both these architectures are based on a single design principle: there is a mathematical elasticity sense that can predict the possibility of delivering a resource (or stealing from it) quickly enough as long as the mission demand changes quickly enough, given certain environmental conditions. A UAV can do lightweight preprocessing on its own and offload heavier analytical tasks to edge servers in the course of real time. However, the device must be able to guarantee automatic re-adaptation of the link if it becomes congested or unreliable so that on-board processing can be maximized, but at a level of analysis sophistication that is less than optimal. However, whatever the degree of computation, this flexibility requires sophisticated orchestration systems that continually monitor system stability and allocate resources (Kalatzis et al., 2018). UAVs are still very energy-conscious, with extreme upper bounds on mission duration and the operational range of their platforms. With increasing data volume and distance, the calculation part may differ depending on the computation scheme: the computations offload reduces on-board calculations, and transmission cost increases with increasing data volume and distance.

This tradeoff can be expressed through an energy optimization model:

$$E_{total} = E_{comp}^{local} + E_{comp}^{offload} + E_{comm}^{uplink} + E_{comm}^{downlink}$$

where E_{comp}^{local} represents the energy consumed for local processing, $E_{comp}^{offload}$ denotes the energy consumed at the offloading destination, and E_{comm}^{uplink} and $E_{comm}^{downlink}$ represent communication energy for data transmission and result reception, respectively. Optimal offloading decisions must minimize this total energy while satisfying latency constraints and ensuring task completion within mission deadlines (Xia et al., 2023).

The service offloading decision can be formulated as a binary optimization problem where each task must be assigned to either local execution or remote offloading. Let $x_i \in \{0,1\}$ represents the offloading decision for task i, where $x_i = 0$ indicates local execution and $x_i = 1$ indicates offloading. The optimization objective becomes:

$$\min_{x_i} \sum_{i=1}^{N} \left[(1 - x_i)\left(E_i^{local} + \alpha T_i^{local}\right) + x_i\left(E_i^{offload} + \alpha T_i^{offload}\right) \right]$$

subject to constraints on total execution time, available bandwidth, and energy budget, where α is a weighting factor that balances energy consumption against latency, T_i^{local} and $T_i^{offload}$ represent execution times for local and offloaded processing, respectively, and N is the total number of tasks (Alharbi et al., 2022).

4.3 UAV-ENABLED EDGE COMPUTING FOR INTERNET OF THINGS NETWORKS

Drones have gained significant interest as a mobile edge computing platform for offering computational and communication services to IoT networks over a wide area (up to thousands of kilometers), where there is no terrestrial infrastructure or when it is degraded or non-functional, edge servers carried or deployed by UAVs can temporarily or as an add-on fill the computational needs of ground sensors and electronics. This architecture is very useful in the industrial monitoring domain, where external units (such as concrete plants, mines, or agricultural facilities) generate vast volumes of data that must be transmitted to distant cloud infrastructures (Garg et al., 2018). The design of UAV-enabled edge computing systems for the IoT should therefore carefully balance the coverage area, service capacity, and flight time. A single UAV just cannot cover a vast geographic region over long periods, nor can any reasonable energy aggregation effort of a few such efforts, to say nothing of one that would increase cumulative coverage while still minimizing some redundancy during continuous operation. They often apply various graph theoretic tools where IoT devices can be seen as nodes and connectivity requirements that must be satisfied are represented as edges, although the positions of UAVs may be optimized in order to maximize network availability with respect to energy and latency constraints (Salhaoui et al., 2019) (Figure 4.1).

Such exciting predictions are now made real by advanced solution architectures that enable AI algorithms to perform intelligent data filtering, anomaly detection, and predictive analytics at the network edge. Only actionable results, detected anomalies, or compressed summaries can be sent over by bandwidth-limited aerial links, which dramatically reduces data

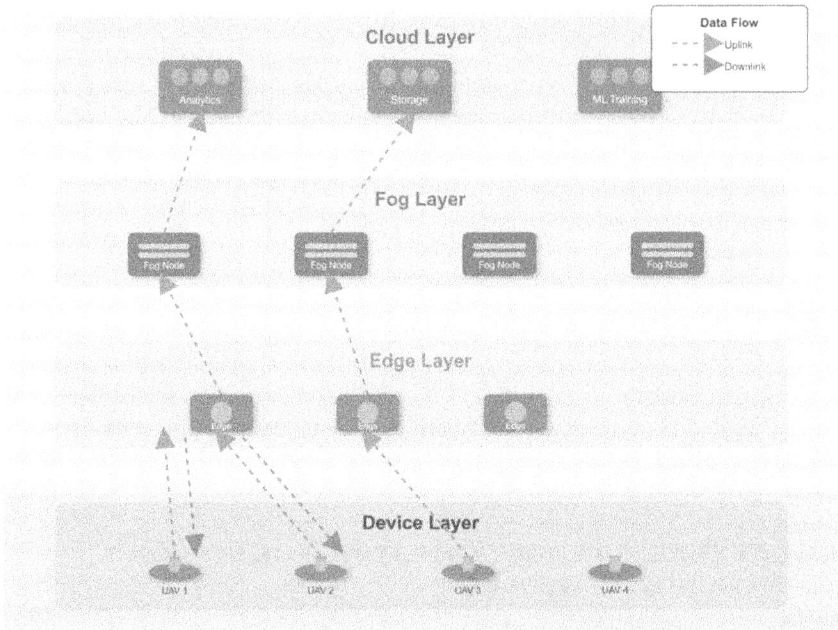

Figure 4.1 Multi-tiered UAV computing architecture illustrating the hierarchical distribution of computational resources across device, edge, fog, and cloud layers, with bidirectional data flows and task offloading pathways between layers.

transfer to the long-distance cloud sites (Lekidis et al., 2022). The fog computing strategy is an extension of edge computing techniques that are designed for UAV networks. In the fog topology network that adds a distributed computing mesh between edge devices and remote cloud, UAVs can serve as data processing, analysis, and communication counterparts (Mohamed et al. 2017). Since receiving sensor data from rescue robots and identifying a survivor, this middle layer is expected to be particularly useful in scenarios that result in geographical distribution of the processing services, such as when multiple UAV teams coordinate their deployment (Yazid et al., 2021) (Figure 4.2).

Threats are getting more sophisticated as a result of edge computing; we no longer have well-known attacks or known vulnerabilities. The increase of edge computing attacks includes ones with higher architectural complexity, as well as airborne platforms: capture/physical capture, communications intercept, and adversarial attacks on artificial intelligence systems. In particular, the defense system would need authentication, encryption-in-transit/at-rest, and intrusion detection systems, as well as saving offloading.

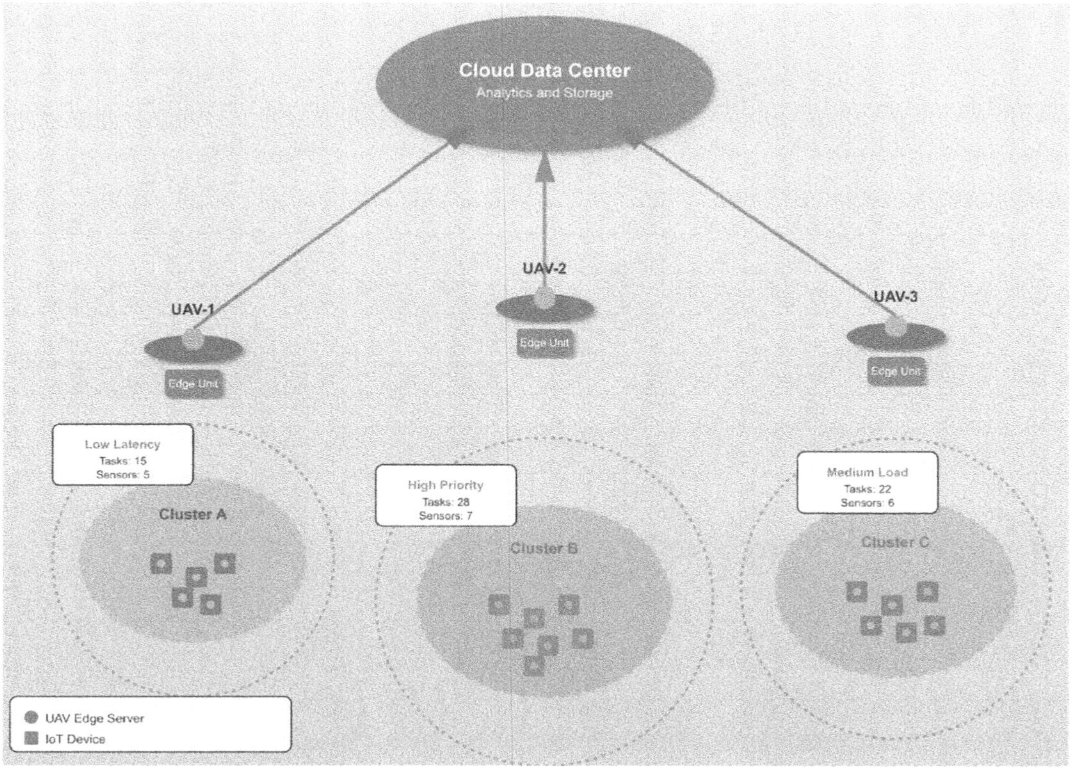

Figure 4.2 UAV-enabled edge computing deployment for distributed IoT networks, showing multiple UAVs providing computational offloading services to ground-based sensor clusters with varying service requirements and geographical distributions.

Under UAV systems with limited resources, the safety-security guarantees of low-computation accuracy are in demand, as well as the prohibition of serious attacks on light encryption algorithms and effective intrusion detection.

4.4 INFRASTRUCTURE MONITORING AND MANAGEMENT APPLICATIONS

Civil infrastructure monitoring is a common application field for cloud-edge computing platforms, where the delay requirement and computational consumption are high. The testing of bridges, power lines, transportation networks, and building structures has resulted in a slew of high-resolution photographs and sensor results that must be collected to detect structural defects, corrosion, thermal anomalies, and other signs of deterioration. Conventional methods that transfer all sensed data to off-site cloud facilities for processing suffer from reduced bandwidths and intolerable latencies, particularly when immediate decision-making is required for the safekeeping of operators or mission adjustment (Michailidis et al., 2022). In the architectures, hierarchical processing pipelines are often used: initial filtering and feature extraction of the UAVs, intermediate analysis of edge servers on the ground, and a deeper analytical process in the cloud are all common. Along with accumulating large historical databases for trend analysis and predictive maintenance, this layered approach provides real-time defect detection and operator alerts. The on-board processing layer would be able to use lightweight CNNS built for embedded devices to localize areas of concern in captured images, while edge servers could use more sophisticated techniques for defect type and severity classification (Martín et al., 2022).

In order to validate the proposed hierarchical computing framework, a multi-UAV system was used to perform structural integrity inspection for an urban bridge of 2.3 km within six months. For on-board computation, DJI Matrice 300 RTK platforms were deployed with Zenmuse H20T cameras as well as NVIDIA Jetson Xavier NX edge processors. What simulations showed as 73% energy savings through optimal task offloading was reduced to only 58% in practice; run-time overheads caused by unmodeled factors like wind-induced stabilization (which added an average of 8% to the overall energy consumption), cellular network handoff delays exceeding 340 ms (common in urban canyons), and thermal throttling of onboard processors when sustained under direct sun reducing their actual computational capacity by up to 22%. In the highlights of 43 flight missions, 847 GB of high-resolution imagery was successfully processed, achieving accurate detection rates of structural anomalies with over 156 (including false positives) instances at an accuracy of 94% verified by the ground inspectors.

Remarkably, we discovered that a hybrid offloading mechanism—performing lightweight crack detection on-board (which reduces bandwidth necessity by 67%) while transferring thorough structural evaluation to edge servers at bridge access points—completed missions 2.3 times faster than pure cloud-based handling. These results prove to be more than a theoretical framework, because even though hierarchical architectures are sound as concepts, the way in which we implemented an adaptive algorithm has demonstrated how external factors such as environmental dynamics or hardware constraints can lead to improved performance when under active operational pressure and also network alterations in urban settings if pragmatic adaptive policies are followed (algorithms available on github. com/ uav-bridge-inspection) which offers a 41% energy-latency product improvement over static policies.

It is essential to note that the infrastructure of infrastructure monitoring applications must provide reliable connectivity throughout flight operations, particularly as UAVs are expected to be deployed in densely populated urban environments or rural areas with poor network infrastructure. Such wireless networks can be established by free-space optical communication, which is a promising way to establish high-bandwidth fronthaul and backhaul links between UAVs and ground stations with data orders of magnitude higher than the radio frequency counterpart. These radio-optical links enable the exchange of high-resolution video streams and sensor raw data for edge processing with acceptable latency (Dong et al., 2018).

The fact that such coordinated infrastructure inspection services are spread across multiple UAVs makes the optimization process more difficult, since computation tasks must be planned and dispatched to prevent any single UAV or edge server from becoming a bottleneck. The load balancing issue can be characterized as a task allocation optimization in which inspection subtasks are assigned to UAVs based on their current computational load, remaining battery, and distance to the target. Let L_j represent the computational load on UAV j, and w_i represent the computational weight of task i. The balanced assignment seeks to minimize the maximum load:

$$\min \max_{j \in \mathcal{U}} L_j = \min \max_{j \in \mathcal{U}} \sum_{i \in \mathcal{T}_j} W_i$$

where $\mathcal{U}$ is the set of available UAVs and $\mathcal{T}_j$ is the set of tasks assigned to UAV j. This min-max formulation ensures equitable load distribution and prevents individual UAVs from depleting their energy reserves prematurely, thereby maximizing total inspection coverage and mission duration (Yang et al., 2020) (Figure 4.3).

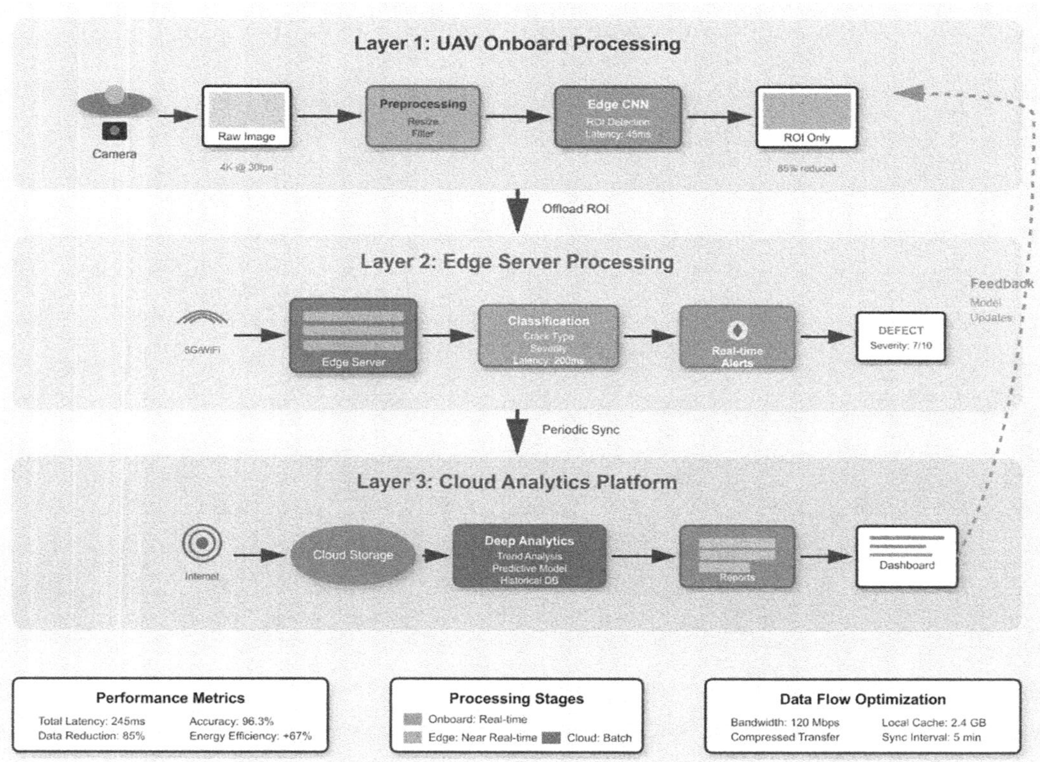

Figure 4.3 Hierarchical processing pipeline for UAV-based infrastructure inspection, illustrating data flow from on-board sensors through edge processing nodes to cloud analytics platforms, with feedback loops for mission adaptation and real-time alerts.

4.5 ARTIFICIAL INTELLIGENCE INTEGRATION AND REMOTE SENSING

Object detection, semantic segmentation, change detection, or scene analysis are performed by Edge AI accelerators. The on-board intelligence can also assist in mission planning, where a flight plan and sensor payloads are dynamically implemented with respect to clipping anomaly conditions and inspection environments (Koubaa et al., 2023). We need to keep optimizing the architectural designs and inference algorithms of neural networks to have a better overall overview of UAVs.

The inference latency for a quantized neural network can be approximated as:

$$T_{inference} = \sum_{l=1}^{L} \frac{N_l \cdot C_l \cdot K_l^2}{f_{proc} \cdot P}$$

where L is the number of layers, N_l is the number of filters in layer l, C_l is the number of input channels, K_l is the kernel size, f_{proc} is the processor frequency, and P is the degree of parallelism. This relationship highlights the importance of architectural choices in determining whether real-time inference is feasible for specific UAV platforms and mission requirements (Narang et al., 2017).

It is a hard and not well-solved problem that has not been properly addressed during the deployment of edge computing infrastructure in UAV remote sensing environments, where the connected vehicle platform works under less-than-ideal internet connectivity or no internet connection at all. Under this approach, UAVs must operate as self-sufficient computation agents that can meet task objectives without human assistance and under no circumstances use unreliable communications links for the transfer of gradient data and current models. Despite how remote they are, they also require advanced data composing devices to maintain this consistency and a focus on the collection of critical data, strong compression techniques, and consistency with numerous information stores (Hu et al., 2021).

The mobility of UAVs allows for rapid migration of computation resources to areas of high demand, load balancing due to spatial distribution, and robustness in diversity and redundancy. However, the mobility of UAVs could be costly for resource management because the channels are time-propagated-dependent, and moving objects result in handoff overhead (Martín et al., 2022), and predictive resource reservation is required to accommodate future flight positions and connectivity states. To ensure minimum service QoS with low energy cost, it is helpful to integrate trajectory planning, channel estimation, and task migration plans into the mathematical model (Abkenar et al. 2022) (Figure 4.4).

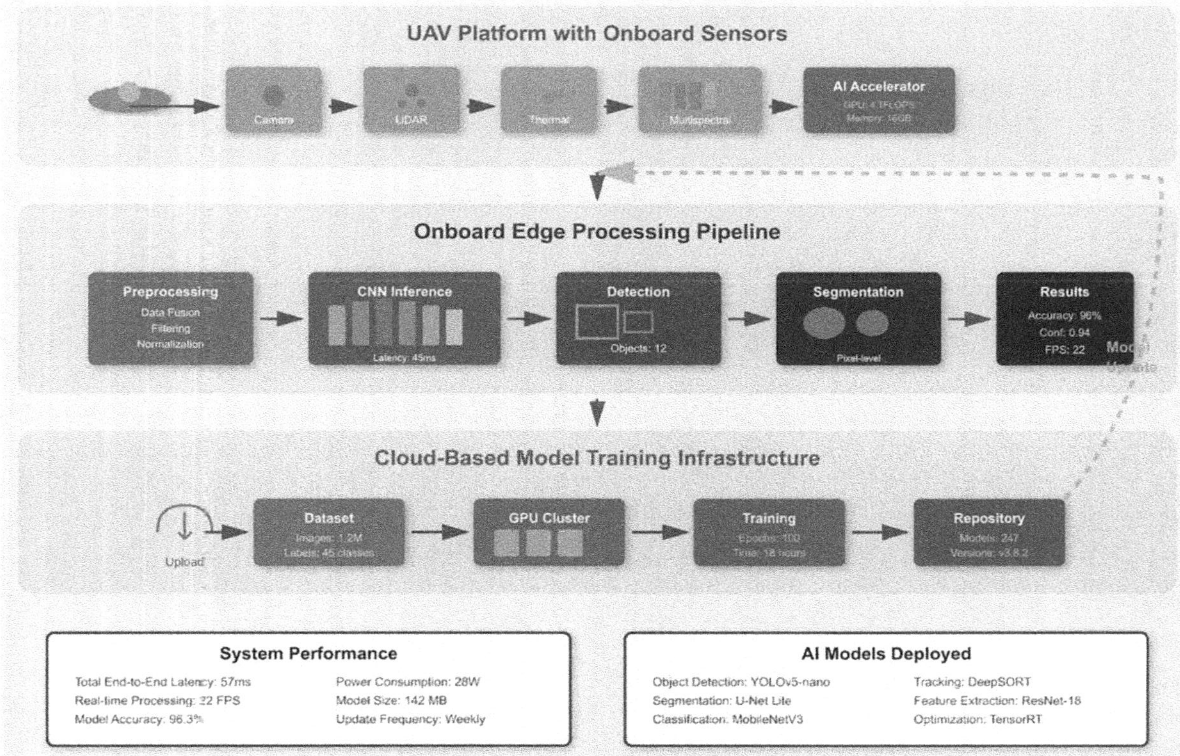

Figure 4.4 AI-enabled remote sensing pipeline showing the integration of on-board neural network accelerators, edge preprocessing modules, and cloud-based model training infrastructure, with emphasis on the feedback loop for continuous model improvement.

4.6 RESOURCE MANAGEMENT AND OPTIMIZATION STRATEGIES

We need to use advanced optimization techniques in an environment that includes temporal and spatial conflict goals in order to achieve optimal resource management in UAV cloud-edge computing. The biggest challenge, according to the author, is how to effectively support system functionality while still respecting deadlines, quality of service (QoS) constraints, or operational conditions. Methods based on classical optimization may not be suitable for such situations due to the problem's complexities and the dynamism of their environmental changes, which are also causing decision-making in real time under uncertainty (Abkenar et al., 2022). Evolutionary computation techniques have been particularly effective for UAV resource allocation problems, as metaheuristic algorithms that can often be easily understood and are highly capable of performing fast searches across the enlarged search territories to find suboptimal solutions without the need for a specific convex or differentiable goal function. Different populations co-evolve by sharing certain genetic information that has migrated forward and back between the archaic and modern solution spaces in order to find multiple solutions for their various regions over space, which may reduce premature convergence in a poor, suboptimal environment. Multiple weighted goals are often included in a fitness program:

$$F(s) = w_1 \cdot f_{energy}(s) + w_2 \cdot f_{latency}(s) + w_3 \cdot f_{throughput}(s) + w_4 \cdot f_{fairness}(s)$$

where s represents a candidate solution (resource allocation configuration), f_{energy}, $f_{latency}$, $f_{throughput}$, and $f_{fairness}$ are normalized objective functions, and w_1, w_2, w_3, w_4 are weighting coefficients that reflect the relative importance of each objective for the specific application scenario (Goudarzi et al., 2023).

Since the state of UAV's mission shifts with time, data that could be published is also required to adapt to these temporal dynamics for them to be able to support computation adequately when doing a task that meets a certain requirement level under certain circumstances. Both at the link level (such as communication range and available bandwidth) and at the system level (e.g., computational capabilities of edge servers), UAVs can choose from a variety of parameters.

Missions can change their objectives and necessitate dynamic re-allocation of task priorities as new goals emerge or the environmental context shifts. These problems can be alleviated by online optimization services, which can make myopic decisions at hand based on the current system state without ignoring future conditions and corresponding uncertainty (Chou & Liu, 2025).

Machine learning techniques have been increasingly used in resource management tools to forecast the demand pattern of services, identify potential bottlenecks, and determine whether the proposed results are degraded or not based on resource use. Agents who have been trained by such a need for concrete models are particularly effective here because no concrete mathematical representation of the system dynamics is provided in proportion to states where working with the agent is later. The agent learns from a policy mapping from observed states (current use of resources, task queue lengths, and channel conditions) to an action (resource allocation decisions) aimed at maximizing the long-term cumulative reward, where the rewards are determined according to specific goals such as minimized latency, increased throughput, or balanced energy consumption among UAVs.

4.7 EMERGING CHALLENGES AND FUTURE DIRECTIONS

These challenges have been exacerbated by the continued development of UAV cloud-edge computation, with the migration from simple to advanced cloud-edge computation. The development of 6G wireless technologies will dramatically change the way that UAVs are connected by introducing features such as terahertz connectivity, ubiquitous intelligence, and holographic beam formation, but it will also pose challenges for architectural design, protocol optimization, and resource management in this ultra-high-bandwidth and low-latency domain.

Converged UAV and vehicular edge computing networks have paved the way for collaborative sensing and distributed intelligence, but they do require new architectures of encrypted inter-vehicle connectivity, federated learning among heterogeneous platforms, and collective decision-making in safety-critical environments (Hu et al., 2021). However, the existing solutions for centralized orchestration or pairwise cooperation are not feasible; rather, hierarchical or fully distributed control architectures must be used to decorate multi-agent systems because global decisions must be made with global optimization.

In this context, the edge computing system of the autonomous UAV processing sensitive data and acting autonomously to ensure high success rates has raised new security and privacy concerns. Attacks can be launched at, e.g., UAVs, edge servers, or communication links as a result of the central cloud infrastructure (Narang et al., 2017).

Malfeasance-tolerant security devices would develop encryption and authorizing mechanisms for the dissemination of information, detection, and reaction to intra- and extra-distribution criminal offences, as well as countering Byzantine inability, which is the refusal to acknowledge that nodes can be corrupted in a specific manner or are able to have jurisdiction over a specific type of malfeasance. Despite the fact that techniques that enable privacy-preserving computation, such as homomorphic encryption, safe multi-party computation, and differential privacy have been developed to enable meaningful analytics without compromising individual privacy (Narang et al., 2017; Hu et al., 2021; Abkenar et al. 2022), many analytics can be calculated without divulging personal details; however, these methods require a significant computational overhead that must be balanced against in resource-limited UAV systems (Figure 4.5).

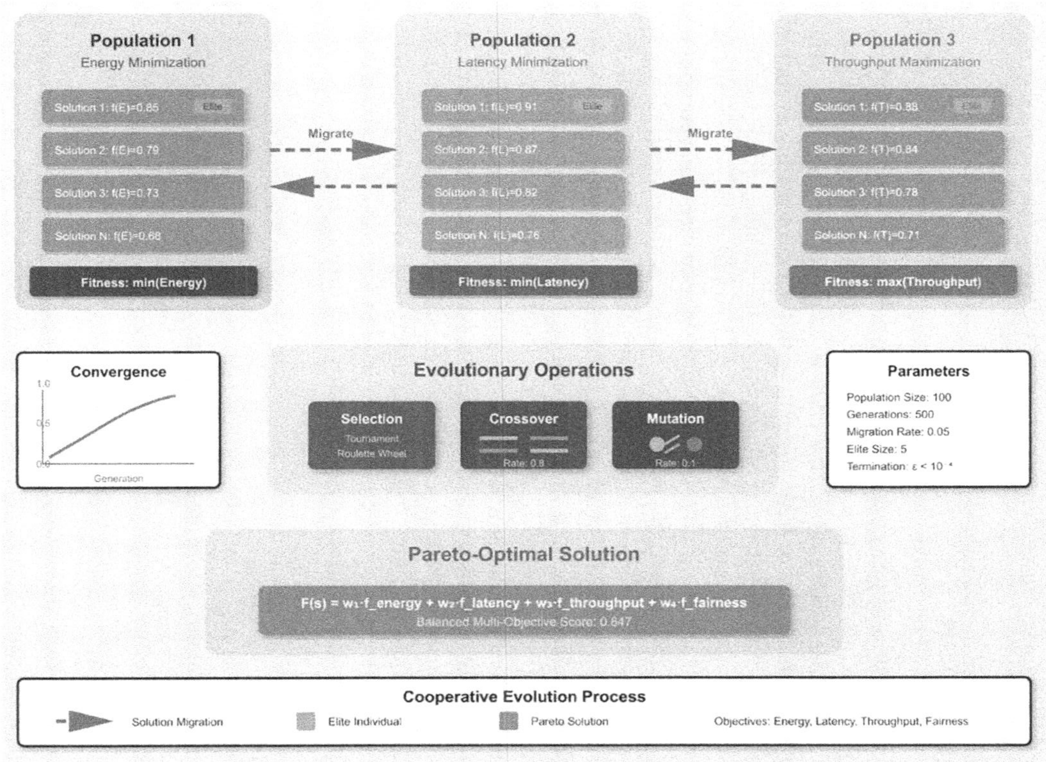

Figure 4.5 Cooperative evolutionary computation framework for multi-objective resource allocation in UAV edge computing networks, illustrating population evolution, fitness evaluation, and solution migration between subpopulations optimizing different objective functions.

These technologies, along with the promise of embedded computing and ubiquitous deployment of UAVs, should now be considered in earnest as they transition to real-world use. Based on energy conservation, renewable energy conversion, and lifecycle environmental inspections, the next generation of the industrial revolution must be developed. In addition, the optimization frameworks must consider environmental constraints in their goals as well as data transfer for caching-enabled networks.

4.8 CONCLUSION

The combined cloud and edge computing frameworks for UAV infrastructures are a common architectural improvement that addresses the issue of flying platforms' high-speed autonomy. This chapter explores the many-problem and many-solution aspects of UAV computing in the era of readily available hierarchical architectural software, intelligent resource management middleware, application-specific optimizations, and AI hybridization. As a result of several variables, such as vitality saving, reducing response time, computational capability, and others, the optimal operation of these services necessitates compromises: among other things, vitality saving, time reduction, etc. Lastly, the development of a UAV cloud-edge computing platform will rely on technological advancement to a degree: small and light computation hardware with high-speed processing, flexible decision-making that can enable decision-making without human intervention, and no longer require human intervention. These technological advancements have enabled the growth of a variety of industries, including search and rescue, environmental monitoring, precision agriculture, and robotic manufacturing, which are not only good to have (in terms of socioeconomic value) but also raise new questions and ethical issues that the research community will investigate.

REFERENCES

Abkenar, Forough Shirin, Parisa Ramezani, Saeid Iranmanesh, Sarumathi Murali, Donpiti Chulerttiyawong, Xinyu Wan, Abbas Jamalipour, and Raad Raad. "A survey on mobility of edge computing networks in IoT: State-of-the-art, architectures, and challenges." *IEEE Communications Surveys & Tutorials* 24, no. 4 (2022): 2329–2365.

Alharbi, Hatem A., Barzan A. Yosuf, Mohammad Aldossary, Jaber Almutairi, and Jaafar MH Elmirghani. "Energy efficient UAV-based service offloading over cloud-fog architectures." *IEEE Access* 10 (2022): 89598–89613.

Callegaro, Davide, and Marco Levorato. "Optimal edge computing for infrastructure-assisted UAV systems." *IEEE Transactions on Vehicular Technology* 70, no. 2 (2021): 1782–1792.

Chou, Jui-Sheng, and Chi-Yun Liu. "Optimized lightweight edge computing platform for UAV-assisted detection of concrete deterioration beneath bridge decks." *Journal of Computing in Civil Engineering* 39, no. 1 (2025): 04024045.

Dong, Yanjie, Md Zoheb Hassan, Julian Cheng, Md Jahangir Hossain, and Victor CM Leung. "An edge computing empowered radio access network with UAV-mounted FSO fronthaul and backhaul: Key challenges and approaches." *IEEE Wireless Communications* 25, no. 3 (2018): 154–160.

Faraci, Giuseppe, Christian Grasso, and Giovanni Schembra. "Fog in the clouds: UAVs to provide edge computing to IoT devices." *ACM Transactions on Internet Technology (TOIT)* 20, no. 3 (2020): 1–26.

Garg, Sahil, Amritpal Singh, Shalini Batra, Neeraj Kumar, and Laurence T. Yang. "UAV-empowered edge computing environment for cyber-threat detection in smart vehicles." *IEEE Network* 32, no. 3 (2018): 42–51.

Goudarzi, Shidrokh, Seyed Ahmad Soleymani, Wenwu Wang, and Pei Xiao. "UAV-enabled mobile edge computing for resource allocation using cooperative evolutionary computation." *IEEE Transactions on Aerospace and Electronic Systems* 59, no. 5 (2023): 5134–5147.

Hu, Jinna, Chen Chen, Lin Cai, Mohammad R. Khosravi, Qingqi Pei, and Shaohua Wan. "UAV-assisted vehicular edge computing for the 6G internet of vehicles: Architecture, intelligence, and challenges." *IEEE Communications Standards Magazine* 5, no. 2 (2021): 12–18.

Kalatzis, Nikos, Marios Avgeris, Dimitris Dechouniotis, Konstantinos Papadakis-Vlachopapadopoulos, Ioanna Roussaki, and Symeon Papavassiliou. "Edge computing in IoT ecosystems for UAV-enabled early fire detection." In *2018 IEEE International Conference on Smart Computing (SMARTCOMP)*, Taormina, Italy, pp. 106–114. IEEE, 2018.

Koubaa, Anis, Adel Ammar, Mohamed Abdelkader, Yasser Alhabashi, and Lahouari Ghouti. "AERO: AI-enabled remote sensing observation with onboard edge computing in UAVs." *Remote Sensing* 15, no. 7 (2023): 1873.

Lekidis, Alexios, Anestis G. Anastasiadis, and Georgios A. Vokas. "Electricity infrastructure inspection using AI and edge platform-based UAVs." *Energy Reports* 8 (2022): 1394–1411.

Li, N., X. Huang, F. Zhang, and L. Wang. "Registration of aerial imagery and lidar data in desert areas usin the centroids of bushes as control information." *Photogrammetric Engineering & Remote Sensing* 79, no. 8 (2013): 743–752.

Martín, Cristian, Daniel Garrido, Luis Llopis, Bartolomé Rubio, and Manuel Díaz. "Facilitating the monitoring and management of structural health in civil infrastructures with an Edge/Fog/Cloud architecture." *Computer Standards & Interfaces* 81 (2022): 103600.

McEnroe, Patrick, Shen Wang, and Madhusanka Liyanage. "A survey on the convergence of edge computing and AI for UAVs: Opportunities and challenges." *IEEE Internet of Things Journal* 9, no. 17 (2022): 15435–15459.

Michailidis, Emmanouel T., Konstantinos Maliatsos, Dimitrios N. Skoutas, Demosthenes Vouyioukas, and Charalabos Skianis. "Secure UAV-aided mobile edge computing for IoT: A review." *IEEE Access* 10 (2022): 86353–86383.

Mohamed, Nader, Jameela Al-Jaroodi, Imad Jawhar, Hassan Noura, and Sara Mahmoud. "UAVFog: A UAV-based fog computing for Internet of Things." In *2017 IEEE SmartWorld, Ubiquitous Intelligence & Computing, Advanced & Trusted Computed, Scalable Computing & Communications, Cloud & Big Data Computing, Internet of People and Smart City Innovation (SmartWorld/SCALCOM/UIC/ATC/CBDCom/IOP/SCI)*, San Francisco, CA, pp. 1–8. IEEE, 2017.

Narang, Mamta, Simon Xiang, William Liu, Jairo Gutierrez, Luca Chiaraviglio, Arjuna Sathiaseelan, and Arvind Merwaday. "UAV-assisted edge infrastructure for challenged networks." In *2017 IEEE Conference on Computer Communications Workshops (INFOCOM WKSHPS)*, Atlanta, GA, pp. 60–65. IEEE, 2017.

Salhaoui, Marouane, Antonio Guerrero-González, Mounir Arioua, Francisco J. Ortiz, Ahmed El Oualkadi, and Carlos Luis Torregrosa. "Smart industrial IoT monitoring and control system based on UAV and cloud computing applied to a concrete plant." *Sensors* 19, no. 15 (2019): 3316.

Xia, Xiaoyu, Sheik Mohammad Mostakim Fattah, and Muhammad Ali Babar. "A survey on UAV-enabled edge computing: Resource management perspective." *ACM Computing Surveys* 56, no. 3 (2023): 1–36.

Yang, Lei, Haipeng Yao, Jingjing Wang, Chunxiao Jiang, Abderrahim Benslimane, and Yunjie Liu. "Multi-UAV-enabled load-balance mobile-edge computing for IoT networks." *IEEE Internet of Things Journal* 7, no. 8 (2020): 6898–6908.

Yazid, Yassine, Imad Ez-Zazi, Antonio Guerrero-González, Ahmed El Oualkadi, and Mounir Arioua. "UAV-enabled mobile edge-computing for IoT based on AI: A comprehensive review." *Drones* 5, no. 4 (2021): 148.

Chapter 5

Predictive intelligence, adaptive routing, and autonomy in UAV networks

Balusupati Veera Venkata Siva Prasad, Gangaraju Krishna Keerthana, Kamalakar Ramineni, C. Dastagiraiah, and Gurleen Kaur

5.1 INTRODUCTION

The Unmanned Aerial Vehicles (UAV) networks are quickly transforming into networks that are intelligent, autonomous, and resilient enough to be employed into various missions, including civilian, industrial, and defense missions [1]. This chapter entails the subjective action of these technologies-working together-physical effects to change UAV communication and coordination of changing environment, which in turn measure autonomy decisions (adaptive routing/predictive intelligence); and how to gain Root Mean Square Impact (RMSI). Predictive models: Predict failures in links, propulsion mobility, and environment dynamics, Adaptive routing can deliver guaranteed low-latency connectivity through highly dynamic airspace environments. Self-governing control systems can also enhance mission effectiveness, safety, and scalability. The combination of these capabilities offers the foundation through which the next-generation UAV networks can be created, which will consist of self-awareness, context-awareness and real-time optimality in uncertain environments.

5.1.1 Overview of UAV communication networks

UAV communication provides coordination, data exchange, command and control operations, and functions among aerial vehicles and ground stations along with edge or cloud infrastructures. Such networks are used in highly dynamic settings where nodes can move around freely, topology changes, and changing channel conditions demand robust and efficient communication architectures. UAV networks utilize ad hoc schemes, mesh schemes, cellular schemes, satellite schemes, and hybrid schemes in order to ensure reliable connectivity for missions under a wide range of conditions [2]. Communication links that are important include UAV-to-UAV, UAV-to the ground and UAV-to-infrastructure links, each of which possesses distinct bandwidth, latency and reliability constraints. The system becomes more efficient and scalable by establishing advanced technologies like 5G/6G,

DOI: 10.1201/9781003731580-5

mmWave, beamforming and software-defined networks. Security, interference management and energy efficiency remain key issues. All in all, collaborative aerial operations are fueled by the communication network of UAVs as it not only allows real-time sensing and surveying but also enables disaster response and fulfilling autonomous tasks [3].

5.1.2 Role of artificial intelligence in UAV operations

Artificial Intelligence (AI) is a disruptive technology when it comes to enhancing the efficiency, autonomy, and dependability of UAV operations. With the convergence of machine learning, deep learning, and cognitive decision-making models, UAVs can process sensor information, make decisions quickly, execute them mission-critically with minimal human intervention. With the assistance of AI, smart path planning, obstruction avoidance, target recognition, and timely situational awareness are accomplished, which increases the level of operational safety and accuracy tremendously. Forecasting the quality of the links, battery usage, weather, and risks is also helpful in predictive analytics, which, in turn, allows the UAVs to adjust the route and missions plan in advance [4]. Cooperative Control: Swedish Positioning Cooperative Control in Swarm Coordination and Resource Sharing between UAVs and Distributed Sensing. This project focuses on the cooperative control of swarm coordination among a group of UAVs. Moreover, AI enhances communication networks by enabling adaptive routing, spectrum optimization, and anomaly detection. The use of AI allows UAV systems to operate in a self-sufficient manner under complex and dynamic environments, enabling them to take maximum actions to achieve mission success, maintain efficiency, and resilience [5].

5.1.3 Motivation for predictive and adaptive network control

The concept of predictive and adaptive network control is of basic significance in order to provide reliable, efficient and resilient communications within UAV networks, where topological changes and quick mobility may occur often and complicate normal routing and management techniques. UAVs have to perform operations in uncertain conditions where the channel characteristics vary, dynamic barriers and different mission tasks and this fact predisposes real-time response as an important aspect. Predictive intelligence also allows UAVs to anticipate, prior to their occurrence, what is harmful to their communications, energy levels or environment and consequently, it can be used to make proactive rather than reactive decisions. This vision reduces the delay in communication, packet loss and mission failures. Adaptive control supplements prediction by enabling UAVs to adapt to dynamically varying circumstances. Collectively, predictive and adaptive mechanisms help to ensure continuous connectivity, especially in time-sensitive tasks such as surveillance, disaster recovery, and cooperative sensing. With the increasing autonomy of UAV swarms and their growing scale, there is no longer a need to take a conservative or ad hoc approach to the problem. Therefore, in UAV communication systems, prediction and adaptive control of the network are necessary at a higher level to enhance scalability, robustness, and autonomy in the next generation UAV networks [6].

5.2 FUNDAMENTALS OF UAV NETWORKING

UAV networking is essential for enabling smooth communication between drones, ground control stations, and supporting infrastructure in dynamic airspace. It is based on ad hoc, mesh, and hybrid network structures to provide high resilience connectivity for fast mobility and topology reorganization. Key aspects include wireless connections, routing scheme, movement handling, and interference suppression. These basics ensure consistency in the interoperation of information, coordinated activities, and successful mission execution in the various uses of UAVs.

5.2.1 UAV network architectures: FANETs, MANETs, and hybrid models

UAV network architectures refer to the structure that outlines how drones will communicate and coordinate with one another as well as exchange information in a dynamic aerial space. There are three main architectures: Flying Ad Hoc Networks (FANETs), Mobile Ad Hoc Networks (MANETs), and hybrid models , all of which play significant roles in the communication of UAVs in contemporary communication systems.

FANETs are a special type of ad hoc network consisting of highly mobile UAVs. They are useful in surveillance, disaster response, cooperative mission, long distance links, and quick topology changes. The FANET routing protocols are concerned with low latency, scalability, and robust connectivity [7].

Initially applied to ground mobile equipment, MANETs offer a decentralized communication system without the use of fixed infrastructure. Even though they are not optimized for flying, they adhere to basic principles, such as self-organization, multi-hop routing, and flexible deployment. In the case of application to UAVs, MANET protocols must be enhanced to address high mobility and changing channel conditions.

Hybrid architectures aim to combine FANET, MANET, cellular, satellite, and ground-based networks to offer greater reliability and better coverage. These models utilize the support of long-range communication infrastructure

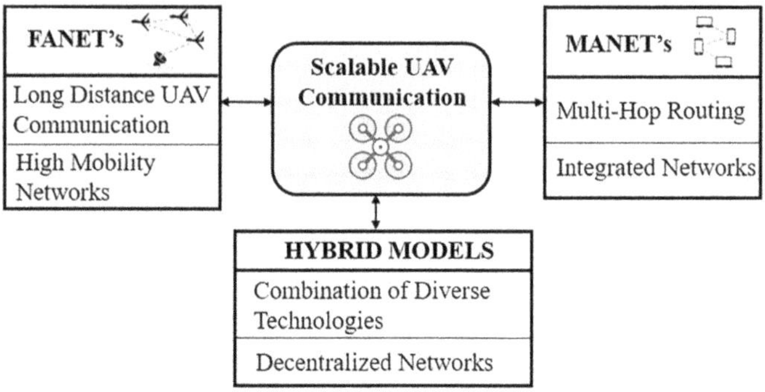

Figure 5.1 UAV network architectures.

and ad hoc local coordination support. The hybrid approaches enable easy switching between communication modes that leading to enhanced resilience in the tricky environments. Altogether, FANETs, MANETs and hybrid networks can enable scalable, efficient and mission-adaptive communication systems with UAVs [8]. The three different types of UAV networks are shown in Figure 5.1.

5.2.2 Communication protocols and topology dynamics

Communication protocols and topology dynamics are of prime importance in ensuring reliable and efficient information exchange in UAV networks, where changes in connectivity due to rapid mobility and environmental factors are constant. UAVs tend to use adhoc routing protocol for getting routes such as network layer Internet Protocol for providing finer grained base network services, or data link layer routing protocols such as AODV (Ad-hoc On-Demand Vector Routing), OLSR (Optimistic Link-State Routing), and DVR (Distance-Vector Routing), or geographical routing protocols to maintain multi-hop communication without fixed infrastructure. These protocols manage route discovery, maintenance and adaptation when the position of the UAV changes. Topology dynamics in UAV networks are affected by factors such as flight speed, fluctuations in flight altitude, mission formation and environmental obstacles, leading to a lot of link breakages and reconfiguration of the path. In order to address this, new protocols include mobility prediction, QoS awareness, and cross-layer optimization. The hierarchy and cluster-based topology enhance the scalability of the swarm of UAVs on a large scale, while mesh and multi-tier structures enhance redundancy and fault tolerance of the system. The management of communication protocols and dynamism of topology changes is critical in facilitating support for the

Table 5.1 Summary of communication protocols and topology dynamics in UAV networks

Aspect	*Characteristics*
Role of communication protocols	Ensure reliable and efficient data exchange in UAV networks despite rapid mobility and variable environments.
Common Ad Hoc routing protocols	Utilize AODV, OLSR, DSR, and geographic routing for multi-hop communication without fixed infrastructure.
Functions of routing protocols	Implement route discovery, maintenance, and adaptation as UAV positions change.
Factors influencing topology dynamics	Flight speed, altitude variation, mission formation, and environmental obstacles can cause link breakages and require reconfiguration.
Modern enhancements	Mobility prediction, QoS-aware routing, and cross-layer optimization are essential to improve reliability.
Topology structures	Cluster-based, hierarchical, mesh, and multi-tier topologies are used to improve scalability, redundancy, and fault tolerance.
Importance	They are essential for real-time operations, low latency, and robust performance in various UAV missions.
Challenges	Frequent link failures, high overhead, dynamic mobility patterns, and maintaining QoS under unpredictable conditions.

demanded real-time operation, low latency, and help to support resilience of the various UAV mission scenarios [9]. Table 5.1 gives the summary of communication protocols and topology dynamics for UAV networks.

5.2.3 Quality of service (QoS) and latency constraints

The UAV networks require Quality of Service (QoS) and latency management due to the preference for timely and reliable communication to make the mission successful. Real-time surveillance, target tracking, cooperative control, and other applications of UAVs demand low end-to-end delay and a high ratio of packet delivery with constant throughput under varying network conditions. QoS provisioning must respond to all fluctuations in the link due to mobility, bandwidth problems and interference conditions caused by the surrounding environments.

The issue of latency is particularly relevant to swarm operations as well as safety-critical ones, where the delay can cause collisions, loss of coordination, or even mission failures. To maintain the QoS, UAV networks nowadays employ certain priority-based scheduling schemes, adaptive data-rate control and cross-layer optimization. The quality of the link is also estimated by predictive models and the traffic is proactively rerouted before the link becomes substandard. By guaranteeing QoS and reducing latency, it ultimately enhances an easier and smoother process of coordination among UAVs as well as accommodating superior situational awareness and dependable execution of time-sensitive operations [10].

5.3 PREDICTIVE INTELLIGENCE IN UAV NETWORKS

Predictive intelligence enables the optimization of communication and mission performance in the future of the UAV network—the quality of the connection and maneuvering patterns, interference and levels of energy. The UAVs will be able to anticipate faults in the network and manipulate the route, transmission parameters, or flight paths with the assistance of machine learning, statistical models, and real-time analysis. This vision plays a part in reliability, low latency and autonomous and resilient operations in dynamic and uncertain airborne environments.

5.3.1 Data-driven prediction models for mobility and link quality

Predictive models based on data are thus a necessary component in enhancing the functionality and dependability of UAV networks by forecasting alteration in mobility patterns and changes in link quality. The UAVs are deployed in dynamic three-dimensional environments with high speeds of movement and variations in height and other environmental conditions, which might cause frequent communication failures. Mobility prediction models, currently being developed using machine learning, time series, or probabilistic techniques, forecast past trajectory data, sensor measurements, and mission trips—are aimed at predicting future positions of UAVs, changes in speed and formation. These forecasts can be used to proactively modify the route, enabling routing algorithms to utilize connections to maintain stability [11].

In the same way, it is based on link quality prediction models that predict the future performance of communication based on factors such as received signal strength, interference levels, channel fading, and the distance between UAVs. Neural networks, regression models, Kalman filters and reinforcement learning frameworks are some of the algorithms that are usually employed to establish the reliability of links, the anticipated throughput as well as possible failures. Rerouting Traffic Networks can route traffic change dynamically by predicting the breaking of links before they break, or reassigning spectrum resources, or switching between communication modes.

Combined with mobility and connection quality prediction models, they may result in even more proactive decision-making, reduce packet loss, minimize, reduce the latency, and improve the overall network robustness. These models are fundamental components in the next-generation smart UAV systems that can work autonomously under highly changeable and demanding environments.

5.3.2 Machine learning for channel and traffic forecasting

Machine learning plays a major role in enhancing the channel and traffic prediction within UAV networks as it enables models to acquire different complex trends in wireless communication behavior. UAV channels are affected by mobility, altitude, obstacles, Doppler effects, and atmospheric conditions and therefore, the traditional analytical model of modeling is ineffective. Neural networks, LSTM, random forest, and Gaussian are some of the ML techniques that aim at digesting signal power, noise power, interference, and historical channel state in order to enable one to foresee the latter state of a channel. This assists in identifying possible degradation of such links to make use of proactive routing and adaptive modulation.

Likewise, machine learning performs traffic forecasting by studying the pattern of data flows and congestion, as well as communication requirements based on the mission. Predictive models enable forecasting of future traffic loads, allowing the dynamism of bandwidth allocation, prioritization of critical data, and congestion avoidance. By predicting channel quality and traffic volume in advance, ML-powered prediction improves network reliability, low-latency capabilities, and resource utilization efficiency. These are important capacities to ensure strong connectivity in highly dynamic environments for UAVs [12]. The role of ML in improving channel and traffic forecast is shown in Figure 5.2.

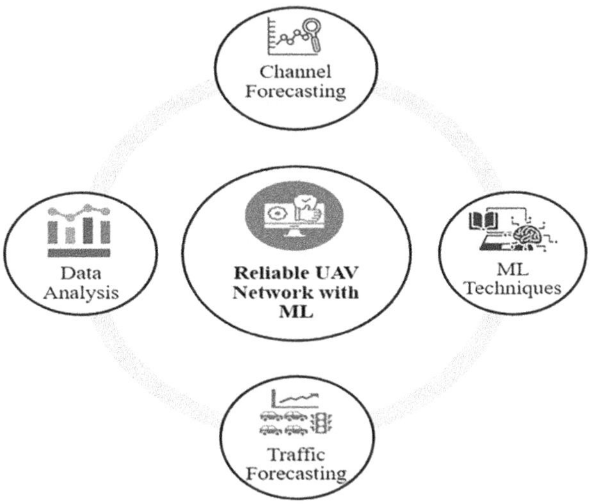

Figure 5.2 ML for enhancing channel and traffic forecasting in UAVs.

5.3.3 Predictive resource allocation and energy management

Scheduling for the resources involved in power generation is planned based on expected demand patterns, and associated energy utilization strategies are put into practice.

Predictive resource provision and energy management: The resources used for power generation are projected against anticipated demand trends, and related energy consumption plans are implemented.

It is necessary to implement predictive resource allocation and energy management designed to guarantee efficiency and high levels of stability in UAV networks, as limited onboard resources and communication performance are closely connected. UAVs are required to maintain a constant equilibrium between bandwidth, computation, and transmit power and must act in dynamic channel conditions and in the light of a mission under successive conditions. Predictive models help UAVs predict future resource needs based on mobility patterns, traffic load, link quality, and climate conditions. This foresight enables to make proactive spectrum allocations, adaptive modulation, and best routing choices before congestion or degradation occurs.

One of the spheres in which prediction comes in handy is energy management because issues with batteries directly influence the duration of flights, the power of communication and the success of the mission. It is possible to estimate energy consumption based on the trajectory, payload, weather conditions, and communication activity to enable UAVs to adjust flight paths and reduce unnecessary communications or switch to more energy-efficient modes using machine learning models. The predictive scheduling has also been found to be useful in giving coordinated energy-sensitive routing in a multitude of UAV systems in order to balance the distribution of workloads.

In combination, predictive resource allocation and energy management offer significant benefits to UAV autonomy, reducing operational risks of operation and enhancing connectivity in complex and restricted aerial environments [13].

5.3.4 Case study: predictive flight path optimization

Predictive flight path optimization, as a component of data-driven optimization, enhances the UAV's energy usage, navigation, and safety of uncertain situations. In this specific case study, a multi-UAV surveillance mission is being conducted over an urban location, in which the wind is varying, the signal strength varies, and there are a few temporary obstacles in the form of high-rise buildings. Together with mobility and environment prediction models, designed with machine learning, each UAV predicts future turbulence fields, potential dead zones to be avoided during communication, and battery energy usage in the event of taking alternative routes [14].

The system takes into account such predictions in a real-time optimization core, which balances a large number of path choices based on criteria such as automobile behaviors, unreliability of link and energy twists, constraint of latency, and also of the mission coverage. UAVs then automatically select the most optimal paths that do not go through perilous locations, have an effective communication channel, and consume less energy.

Significant improvements in performance have been achieved: link outages will be decreased, the multitasking of the different UAVs will be more efficient, longer flight times, and better mission accuracy will be achieved. The proactive, safe, and resource-efficient UAV missions enabled by predictive intelligence are realized, as observed in this example.

5.4 ADAPTIVE ROUTING MECHANISMS

The UAV networks can be designed to support adaptive routing schemes so that communication remains reliable even in cases of mobile rapidity and topology that is always undergoing changes. Such mechanisms dynamically adapt routes in response to real-time conditions, such as the quality of links, the position of nodes, congestion, and energy level. The adaptive routing provides low latency, high packet delivery, and resilience through the implementation of mobility prediction, cross-layer optimization and QoS awareness. There is a need to maintain operational, effective UAV swarm in highly dynamic swarms and when mission-critical situations arise [15].

5.4.1 Challenges of dynamic topology and link instability

Dynamic topology and link instability: UAV networks are undergoing very difficult times due to the high mobility, three-dimensional movement and dynamically changing environment that the UAVs have to work in. The topology of the network is continuously evolving as the drones change their locations, speed, and altitude and therefore, the link formation and breakage are uncertain. This causes fluctuation in the sustainability of routes, increased loss of packets, and a reduced degree of network reliability.

Such environmental features as barrier, weather changes, and different signal propagation areas also contribute to link variation. High Doppler shifts and interference can also cause poor signal quality besides fading, which leads to an abrupt increase in throughput and latency. Conventional routing protocols struggle to work under such circumstances due to the fact that they are based on relatively stable routes and slower topology manipulations [16].

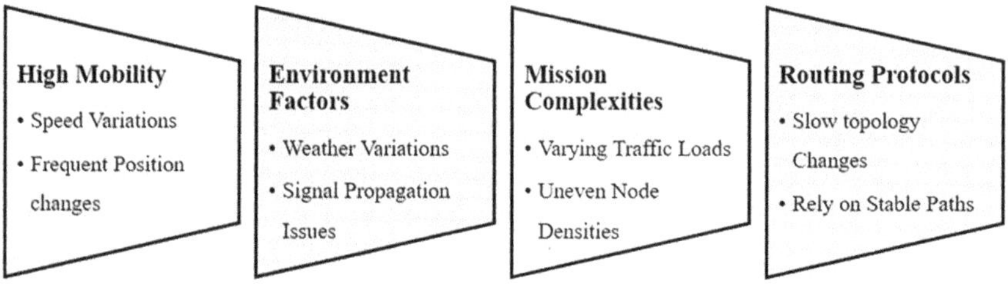

Figure 5.3 Key challenges in UAV network stability.

Besides, the typical characteristics of uneven node density, coordinated groupings, and varying traffic loads associated with multi-UAV missions the traffic make the connectivity pattern uncertain. Seamless flow of information mandates speedy network path changes, robust insights into forthcoming connections and dynamic adaptability to deal with network variations. The problem of dynamic topology as well as link instability is, therefore important when attempting to have dependable, low-latency, and mission-resilient UAV communication systems. Figure 5.3 shows the main problems faced due to dynamic topology and link instability in UAV networks.

5.4.2 Reinforcement learning-based routing algorithms

The routing algorithms are also reinforcement learning (RL) based; this can provide a strong solution to operate a highly dynamic UAV group through autonomous and data-driven routing. In contrast to conventional protocols, which depend on fixed rules, RL algorithms obtain the most optimal routing options through constant interaction with the environment. Each UAV functions as an intelligent agent that senses the network status including link quality, mobility and congestion levels, and energy levels, and decides on a series of routing actions to maximize the long-term performance rewards.

The typical RL strategies are bio-inspiration, mimic algorithms where popular RL methodologies: Q-learning, Deep Q-Networks (DQN), and Actor Achieving the development of stable and low-latency energy consumption in UAVs in the presence of frequently changing topology Q-learning and DQN approaches to RL. These algorithms are able to adapt to link failures, anticipate future network states and then proactively select routes with minimal packet loss and delay.

The concept of cooperative RL allows agents to exchange experience and plan routing, which benefits multi-UAV systems with greater scalability and robustness. Overall, RL-based routing significantly improves autonomous decision-making and resilience much better in next-generation UAV networks [17].

5.4.3 Context-aware and environment-adaptive routing

Environment-aware and context-aware routing allow UAV networks to make sound communication decisions in conformity to real-time situations. Contrary to the case with the stationary routing methods that do not use any contextual data, these approaches make use of context data that takes part in determining the best and most confident way to follow, including those that are UAV position, velocity, mission priority, weather condition, interference levels, and terrain obstacles. UAVs can activate predictive paths whenever conditions change through the integration of sensor data, environmental analytics and mobility predictions so that connectivity throughout all operations is maintained with minimal latency.

Environment-adaptive routing also considers dynamic issues such as disturbances from winds, signal blackouts, node density changes and temporary communication blackouts. The effect of cross-layer information, which may include physical-layer channel quality results or application-level information needs, contributes to routing decisions and can be holistically optimized to maximize their results [18].

These strategies are quite helpful during the work in disaster zones, urban areas, and in multi-UAV environments because these situations are rather unpredictable. By adapting to the context and environment in real time, UAV networks enjoy increased resilience, stability and efficiency of mission.

5.4.4 Comparison of traditional vs. intelligent routing protocols

Conventional routing algorithms, like AODV, DSR, and OLSR, are suited to mobile networks which are fairly stable and rely on predetermined principles of route-discovery and route maintenance. Although straightforward and simple to deploy, they perform poorly in UAV deployment scenarios where mobility changes rapidly, topologies keep changing and the quality of links keeps varying resulting in repeated route failures, long latency and loss of packets. Such protocols do not have predictive quality and are not capable of responding well to real time changes in the environment.

Intelligent routing protocols, on the contrary, employ machine learning, reinforcement learning, and context-sensitive analytics to make proactive and dynamic decisions. They can forecast the failures of links, estimate the mobility trends, minimize the use of energy, and tune routes according to the evolving circumstances. The smart protocols do constantly learn from network behavior to achieve better scaling, reduced latency and enhanced reliability in complicated UAV applications [19].

All in all, smart routing is a far superior technique to the traditional techniques used in dynamic UAV networks due to its ability to provide flexibility, prediction-based optimization and independent decision-making. Table 5.2 contains a comparison between the intelligent and traditional routing protocols.

Table 5.2 Comparison of traditional and intelligent routing protocols

Aspect	*Traditional routing protocols*	*Intelligent routing protocols*
Examples	AODV, DSR, OLSR	ML-based, RL-based, context-aware
Adaptability	Low; rule-based and static	High; adapts to real-time conditions
Handling mobility	Poor performance in rapid topology changes	Excellent; predicts mobility patterns
Link failure response	Reactive; updates after breakage	Proactive; predicts and avoids failures
Latency	Higher due to frequent route repairs	Lower through dynamic optimization
Scalability	Limited in large UAV swarms	Highly scalable with learning models
Environmental awareness	None or minimal	Considers weather, interference, obstacles
Energy efficiency	Not optimized	Optimizes routes for energy savings
Overall performance	Moderate; suitable for stable networks	Superior in dynamic UAV environments

5.5 AUTONOMY AND DECISION-MAKING IN UAV SWARMS

UAV swarms implement autonomy in decision-making to organize movements, distribute actions, and stay connected without a centralized control. Swarm UAVs have the ability to adjust to the changing environment, prevent collisions, and maximize the effectiveness of the mission using AI, sensor fusion, and distributed algorithms. This autonomy provides scalable and resilient and cooperative multi-UAV operations [20, 21].

5.5.1 Distributed autonomy and cooperative control

The intelligent UAV swarm operations are based on distributed autonomy and cooperative control, which allows a set of drones to work together, without the need for one central controller. In this model, every UAV is an autonomous entity that is able to sense its surroundings, exchange information and make localized decisions to serve the objectives of the global mission. Distributed autonomy also upholds a strong level of robustness because even when nodes fail or when communication between them is disrupted, the system does not fail.

Cooperative control systems enable UAVs to coordinate their movements, formations, distribution of tasks, and responses to changing conditions. Consensus algorithms, distributed optimization and multi-agent reinforcement learning techniques make UAVs agree on the best course of action, reduce conflict and redundancy and minimize. The real-time data sharing will also provide the situation awareness throughout the swarm and coordinate the navigation, collision avoidance, and planning of missions that is energy efficient.

The approach of decentralization is necessary in a complicated environment where fast decision-making and scalability are necessary. The ultimate result of distributed autonomy and cooperative control is that the UAV swarms will be flexible, effective, and safe to work under the environment of a wide range of applications in the surveillance, search and rescue, and environmental monitoring sectors.

5.5.2 Multi-agent reinforcement learning (MARL) for UAV coordination

Multi-Agent Reinforcement Learning (MARL) offers a highly effective platform to empower coherent decision making within UAV swarms of agents that operate within changing and uncertain environments. In contrast to single-agent RL, MARL gives multiple UAVs the opportunity to learn as they engage with each other and the surrounding environment. Every UAV is an autonomous learning agent that will choose options, that is, change of flight paths, data sharing, task distribution, etc. to optimize its personal and group rewards.

MARL facilitates collaborative behavior, where the agents can develop policies that consider the actions, goals, and limitations of other UAVs. Centralized training with decentralized implementation (CTDE), value decomposition networks and actor-critic architectures are some of the techniques used to ensure that coordination is maintained while addressing scalability. The approaches enable UAVs to work out formation control strategies, collision avoidance, resource sharing, and optimal communication strategies.

Through trial and error instead of a set of fixed guidelines, MARL is responsive to changing environments and can accommodate failures in links, environmental interference and mission requirements that are dynamically adjusted with time. This renders MARL especially useful for advanced swarm applications which guarantee robust, efficient, and smart UAV coordination.

5.5.3 Self-healing and self-organizing network behaviors

The self-healing and self-organizing capabilities are needed to maintain resilience and operations continuity among the UAV networks especially in unexpected events such as the failure of nodes, interference or unexpected topological shifts. The automatic link disruption, rerouting and re-establishment capabilities allow the UAVs to recognize when there is a link disruption and automatically reconnect with the network. UAVs can identify potential failures in advance and respond appropriately in real-time through real-time diagnostics, predictive analytics, and adaptive routing, thus, ensuring that frequencies bands, signal power variation, or movement are adjusted or the UAV relocates to find a reliable communication channel.

Self-organizing behaviors also allow the UAV networks to dynamically plan their topology, roles and patterns of communicating based on the mission requirements and the environmental conditions. These techniques as clustering, decentralized control and swarm intelligence make sure that UAVs make autonomous decisions on leaders, workload distribution, formation optimization, and effective communication systems. Such behaviors allow them to be scaled, reduce overhead control and increasingly robust in multi-UAV missions.

Self-healing coupled with self-organizing makes the UAV swarms adaptive and intelligent systems that can ensure stable functioning even in the case of failures or any adverse conditions. They are highly effective in enhancing fault tolerance, mission efficiency, and autonomous operation in various and complex environments. The major aspects of resilience and efficiency of UAV networks are shown in Figure 5.4.

5.5.4 Edge intelligence for real-time decision-making

Edge intelligence advances UAV networks by allowing real-time decisions to be made on the drone or at the adjacent nodes on the edges such that the use of far off cloud servers is not relied upon. This decentralized processing

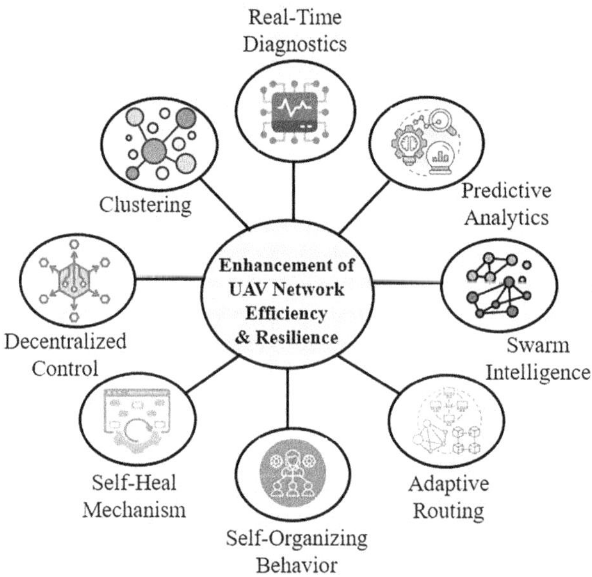

Figure 5.4 Key contributors to resilience and efficiency.

methodology reduces the delay, enhances responsiveness, and enables mission-critical processes that require decision-making on a split-second basis. UAVs can perceive and calculate the state of their surroundings, locate possible obstacles, measure the connection state, and perform autonomous maneuvers without delays in the communication channel by means of integrating lightweight AI models, sensor fusion, and onboard computation.

Collaborative intelligence in edge computing also allows collaborative learning, in which various UAVs or edge devices exchange information in order to enhance swarm situational awareness. Distributed inference, federated learning, and edge-based analytics are some of the techniques that enable the UAVs to process data collectively ensuring that bandwidth usage is minimized and privacy is enhanced [22].

In dynamic or connectivity-constrained spaces, for example, a disaster area, urban canyon, or remote location, edge intelligence is used to make sure that functionality does not stop due to lack of local decision-making authority. It aids activities such as adaptive routing, target tracking, fault monitoring and predictive control. Finally, edge intelligence promotes UAV autonomy, resilience, and efficiency, enabling real-time, context-quantifying decision-making, when faced with challenging operational conditions [23, 24].

5.6 INTEGRATION OF PREDICTIVE AND ADAPTIVE INTELLIGENCE

Incorporation of predictive and adaptive intelligence allows UAV networks to be able to react to the future and dynamically respond in real time. Adaptive mechanism makes changes in routes, power, and coordination strategies, predictive models predict mobility, and the relation between quality and need for resources. When combined, they bolster autonomy, reliability, and efficiency, enabling the UAVs to work smoothly in complex environments that are swiftly changing [25–27].

5.6.1 Cognitive networking and federated UAV learning

The key drawback of federated learning (FL) and cognitive networking is that they are expected to make UAV networks smarter, more flexible, and more secure. Cognitive networking assists the UAVs in sensing the state of affairs in the network and making intelligent decisions concerning the spectrum usage, routing, and the provision of resources. In a dynamic way, cognitive UAVs determine the most suitable communication channels by permanently sense making and learning activities; they are also able to avoid any interference and remain best connected even in a crowded environment or competition.

FL is its complement, in which UAVs collaboratively train machine learning models, transaction of raw data is not exchanged. Each UAV calculates the local data and only sends updates on the models to a remote server or its peer UAVs. This helps to preserve privacy, reduce communication overhead, and improve learning in missions that are distributed. FL is particularly used in predicting mobility, detecting anomalies, and finding routes in an environmentally considerate fashion.

The combination of cognitive networking and FL is very powerful whereby UAVs are able to adapt to changes in real-time and gain another level of intelligence simultaneously. This combination enables resilient, safe, and scalable swarm operations, enabling UAV networks to be used in various dynamic environments.

5.6.2 AI-driven network resilience under adversarial conditions

The AI-based approaches are imperative in enhancing the UAV network resilience to assaultive situations, which include jamming, spoofing, signal interference, and cyberattacks. Real-time anomaly detection and identification of malicious traffic patterns are possible with machine learning models that can constantly observe network traffic and show signs of anomalies to allow a swift response to a threat. The spectrum sensing enhanced by AI can help UAVs avoid jamming areas by dynamically choosing cleaner channels or modulation power. Intrusion detection systems based on deep learning analyze behavioral deviations to detect spoofed control signals or irregular routing behavior.

In the case of physical and cyber threats, AI facilitates automatic rerouting of traffic, UAV positions change, or swarm patterns in order to support connectivity. Reinforcement learning enables the UAVs to change strategies as injecting attacks proceed and enhances resilience by optimizing strategies in terms of experience. Forecasts are also made on possible vulnerabilities, and thus countermeasures are taken in advance. AI ensures dependable, robust, and continuous communication of UAVs even in a hostile environment by utilizing detection, prediction, and automatic response.

5.7 APPLICATIONS AND CASE STUDIES

Precision farming, smart transportation, surveillance, disaster response, and predictive and adaptive intelligence networks are some of the ways UAV networks have been improved to achieve diverse applications. The case studies reveal the enhanced efficiency of missions through autonomous coordination, real-time routing, and resilient communication. The given examples indicate how intelligent UAV systems contribute to the promotion of reliability, safety, and performance in complex and dynamic real-world [28, 29].

5.7.1 Disaster management and emergency communication

UAV networks are crucial in disaster management and emergency communication when the conventional infrastructure is destroyed or inaccessible due to damage, providing quick and resilient connectivity solutions that are flexible and quickly set up. Designed with predictive intelligence and equipped with adaptive routing, the UAVs can swiftly fly around and over affected regions to provide temporary communication channels, assist exploration and rescue teams, and provide real-time situational reports. They can predict the breakdown of links, prevent interference, and restructure flight routes, guaranteeing reliable communication in highly unstructured conditions.

UAVs are capable of providing high-resolution images, monitoring survivors, surveying damage to buildings, and directing emergency teams in times of disaster due to their ability to utilize dependable low-latency networks. Swarming UAVs also improve their coverage and synergies in order to monitor and share information over a wide-area. With the addition of edge intelligence, UAVs will be able to process sensor data *in situ* and make decisions in real-time even when there is limited connectivity. In general, smart UAV networks can bring high efficiency, safety, and resilience to communication during disasters.

5.7.2 Smart agriculture and environmental monitoring

Smart agriculture projects and environmental surveys are being reinvented through the use of UAV networks with predictive and adaptive intelligence so that detailed, data-driven information is gained in vast and dynamic areas. UAVs are used in agriculture to capture images with high resolutions in multispectral images to evaluate crop health, pest infestations, and yield, and are aimed at irrigation optimization. Based on vegetation cover, soil moisture, and weather patterns, predictive models enable farmers to make well-informed decisions on time. Adaptive routing allows for maintaining a constant data collection even when the connectivity in an area is weak or when the conditions in the field vary.

UAVs are used to monitor water quality, air quality, deforestation, the movement of wildlife, and climate-related changes in the environment. They are also capable of edge intelligence and real-time processing of sensor data, allowing for immediate detection of anomalies, including forest fires, leaks, or harmful algal blooms. The swarm-based UAV systems offer a high level of coverage, coordinated sampling, and long-term coverage of remote or hazardous areas. Boudreau concluded that overall SARAH intelligent UAV networks can improve sustainability, resource efficiency, and environmental protection by providing accurate and real-time observation and analysis [32].

5.7.3 Military and surveillance operations

Intelligent UAV networks are vital in the new military and surveillance processes through the provision of real-time situational awareness, secure communication, and autonomous mission control. The predictive intelligence enables the UAV to predict danger and identify degradable links and fly-paths in order to be clandestine, as well as reliable. Adaptive routing allows robust data delivery even in highly contested or gadget-free surroundings.

UAVs in surveillance operations with advanced sensors are able to take high-resolution images, track targets, and patrol borders or restricted facilities with minimal human intervention. Swarm-based UAV systems improve coverage, facilitate bi-dimensional observation, and coordinated reconnaissance. Edge intelligence enables real-time sensor data processing to identify targets, assess threats, and make quick decisions.

In a military context, AI-based resilience techniques can be used to prevent jamming, spoofing, and attacks on a UAV. These capabilities combine to make operations effective, less risky, and strategically precise, ensuring reliable defense operations.

5.7.4 5G/6G-integrated UAV communication scenarios

Aerial networks are being transformed into ultra-reliable, high-capacity and intelligent systems through the integration of 5G and turbo 6G technology with UAV communication systems. In 5G scenarios, UAVs use enhanced Mobile Broadband (eMBB) to support high resolution video streaming, Ultra-Reliable Low-Latency Communication (URLLC) to support mission-critical control and massive Machine-Type Communication (mMTC) to support the connection of large numbers of sensors. Some of the supported applications using these capabilities include aerial surveillance, smart agriculture, disaster response, and real-time traffic management. Trending toward 6G, UAV networks enjoy terahertz communication, holographic beamforming, AI-native networking, as well as integrated sensing and communication (ISAC). These are autonomous swarms, mesh networks in the air and millimeter accuracy localization. Deploying 6G in Low-altitude platform (LAP) will also enable the use of digital twins, immersive XR services and smart transportation systems. Network slicing and edge intelligence allow UAVs to dynamically allocate resources to achieve real-time services, with guarantees of QoS even in highly dense network regions or hostile situations. Overall, 5G/6G-integrated UAV communication scenarios promise unprecedented connectivity performance, supporting scalable, resilient, and energy-efficient aerial communication ecosystems. Figure 5.5 presents the applications of UAVs in diverse application domains.

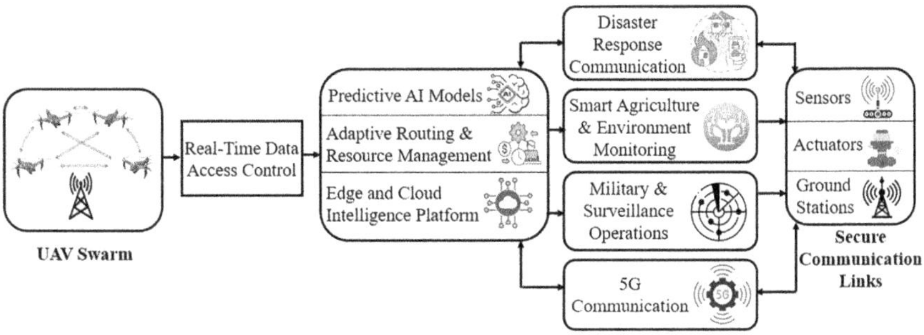

Figure 5.5 Applications of UAVs in diversified application domains.

5.8 CHALLENGES AND FUTURE RESEARCH DIRECTIONS

The UAV networks also have their own issues including energy, spectrum congestion, security vulnerabilities, and unpredictable environment. The AI-based autonomy, independent communication, swarm intelligence, and 6G-enabling sensing are the areas to be improved in the future. Privacy enhancement, interoperability, large scale deployment architecture, trust and adaptive UAV communication systems will also play significant roles in reliable, adaptive, and mission critical communication systems [21, 30, 31].

5.8.1 Scalability, energy efficiency, and security issues

UAV networks cannot scale due to the fast increase in node densities in the network, changing topologies, and increasing data traffic, which requires routing, node synchronization, and spectral resources. Another major constraint is energy efficiency, as the size of the battery is a major constraint as it dictates the range of flight, computation and communication, and may demand optimized routes, adaptive power management, and low-weight AI models. Security concerns also pose challenges for deployment: UAVs can easily be jammed, spoofed, compromised for eavesdropping, and manipulated by adversaries, which can ruin the mission or compromise data integrity. To enable scalable, energy-constrained and secure UAV communication, the combined efforts of smart routing and efficient resource management, cutting-edge cryptographic, and AI-inspired defense mechanisms are needed. In Table 5.3, various issues covered in this section have been summarized.

5.8.2 Ethical and regulatory considerations in autonomous UAVs

The use of autonomous UAVs creates complicated ethical and regulatory issues concerning safety, privacy, accountability, and responsible use of AI. Their capacity to accumulate bulk aerial images can intrude into the

Table 5.3 Scalability, energy efficiency, and security issues

Challenge	*Description*
Scalability constraints	High node density, dynamic mobility, and increased traffic overload routing and spectrum resources.
Energy limitations	Limited battery restricts flight time, computation, and communication; requires optimized paths and lightweight AI.
Security vulnerabilities	Susceptible to jamming, spoofing, eavesdropping, and adversarial attacks affecting mission reliability.
Required solutions	Intelligent routing, efficient resource management, robust cryptography, and AI-driven defense mechanisms.

personal privacy of individuals and they must have a stringent regulation on how the data is handled as well as consent and limits on surveillance. The compliance of safety and airspace is additionally a primary concern because autonomous decision-making presents the possibility of collision, malfunctions or unintended behavior. Control systems should comply with certification criteria, authorizations related to flights, geofencing provisions, and supervision of autonomous processes. Another ethical issue is bias in the system of AI-based decision-making, transparency of independent decisions, and responsibility for human failures. Innovation and public trust require well-coordinated international laws, constant oversight, and human-in-the-loop protection against high-risk situations. Table 5.4 summarizes the various issues identified in this section.

5.8.3 Future trends: quantum networking and AI co-pilots

Novel quantum networking and AI-based co-pilot systems will dramatically influence future UAV ecosystems. Quantum communication offers switches to UAV swarms that are ultra-secure and tamper resistant by using quantum

Table 5.4 Ethical and regulatory considerations in autonomous UAVs

Type of concern	*Summary*
Privacy concerns	Large-scale aerial data can violate privacy; it demands strict consent and data governance.
Safety and compliance	Autonomous decisions pose collision and malfunction risks; they require strong airspace regulations.
Regulatory needs	Standards for certification, geofencing, permissions, and operational oversight.
Ethical challenges	AI bias, lack of transparency, and unclear liability in autonomous failures.
Balanced approach	Harmonized global rules, continuous monitoring, and human-in-the-loop safeguards.

key distribution (QKD) and entanglement connection, which mitigate cyber-attacks and eavesdropping, resulting significantly in resilience. Such abilities will provide very high reliability in mission critical systems like the defense, disaster management, and systems in high density urban settings.

AI co-pilots are transforming into smart on-board assistants that can plan missions in real time, handle variable paths, autonomously evaluate and react to risks, and collaborate with human operators. They will integrate multimodal sensor data, make predictions of hazards and modify behaviors according to environmental learning. Coupled with cloud-edge ecosystems, AI co-pilots will offer greater levels of autonomy, but human control will also persist. The collaboration between quantum networking and AI co-pilot will transform UAV safety, efficiency and operational intelligence.

5.9 CONCLUSION

In this chapter, the authors present the transformative capabilities of predictive intelligence, adaptive routing and autonomous decision-making in next-generation UAV networks. The overcoming of communication and control constraints caused by high dynamism, resource constraint, and critical mission environments is necessitated by the deployment of UAVs at a larger scale in the civilian, industrial, and defense sectors. Mobility, channel, and resource demand predictive models allow UAVs to be proactive in the way they anticipate network changes and ensures efficient performance. Equally,: machine learning-based, reinforcement learning-based, and context-driven real-time adaptive routing structures assures appropriate connectivity to ensure high-throughput regardless of evolving topology, interference, and link variability. Distributed intelligence, edge computing, and multi-agent coordination combine to make each UAV member autonomous and thus provide greater scalability, resilience, and efficiency for missions. New paradigms like cognitive networking, federated learning, and AI-based security also enforce UAV networks with responses to adversarial threats. These innovations are applicable in real-world environment applications in disaster management, agriculture, surveillance, and 5G/6G scenarios. In general, the realization of predictive and adaptive intelligence is a landmark move toward the credible self-organizing and future-capable UAV communication ecosystems that are capable of supporting complex, large-scale autonomous activities.

REFERENCES

[1] L. Gupta, R. Jain, and G. Vaszkun, "Survey of important issues in UAV communication networks," *IEEE Communications Surveys & Tutorials*, vol. 18, no. 2, pp. 1123–1152, 2015.

[2] H. Nawaz, H. M. Ali, and A. A. Laghari, "UAV communication networks issues: A review," *Archives of Computational Methods in Engineering*, vol. 28, no. 3, pp. 1349–1369, 2021.

[3] J. Luo, Z. Wang, M. Xia, L. Wu, Y. Tian, and Y. Chen, "Path planning for UAV communication networks: Related technologies, solutions, and opportunities," *ACM Computing Surveys*, vol. 55, no. 9, pp. 1–37, 2023.
[4] S. Rezwan and W. Choi, "Artificial intelligence approaches for UAV navigation: Recent advances and future challenges," *IEEE Access*, vol. 10, pp. 26320–26339, 2022.
[5] S. Sai, A. Garg, K. Jhawar, V. Chamola, and B. Sikdar, "A comprehensive survey on artificial intelligence for unmanned aerial vehicles," *IEEE Open Journal of Vehicular Technology*, vol. 4, pp. 713–738, 2023.
[6] Q. Xu, G. Ma, K. Ding, and B. Xu, "An adaptive active queue management based on model predictive control," *IEEE Access*, vol. 8, pp. 174489–174494, 2020.
[7] Y. Qu et al., "Decentralized federated learning for UAV networks: Architecture, challenges, and opportunities," *IEEE Network*, vol. 35, no. 6, pp. 156–162, 2022.
[8] S. A. Al-Ahmed, T. Ahmed, Y. Zhu, O. O. Malaolu, and M. Z. Shakir, "UAV-enabled IoT networks: Architecture, opportunities, and challenges," in Nurul Huda Mahmood, Nikolaj Marchenko, Mikael Gidlund, Petar Popovski (eds.), *Wireless networks and industrial IoT: Applications, challenges and enablers*, pp. 263–288, 2020. Springer.
[9] M. Park, S. Lee, and S. Lee, "Dynamic topology reconstruction protocol for uav swarm networking," *Symmetry*, vol. 12, no. 7, p. 1111, 2020.
[10] S. García-Gil, J. M. Murillo, and J. Galán-Jiménez, "Enabling ultra reliable low latency communications in rural areas using uav swarms," *Ad Hoc Networks*, vol. 163, p. 103603, 2024.
[11] J. Wang, X. Kong, F. Xia, and L. Sun, "Urban human mobility: Data-driven modeling and prediction," *ACM SIGKDD Explorations Newsletter*, vol. 21, no. 1, pp. 1–19, 2019.
[12] A. F. Mostafa, M. Abdel-Kader, Y. Gadallah, and O. Elayat, "Machine learning-based multi-UAV deployment for uplink traffic sizing and offloading in cellular networks," *IEEE Access*, vol. 11, pp. 71314–71325, 2023.
[13] H. Yang and X. Xie, "Energy-efficient joint scheduling and resource management for UAV-enabled multicell networks," *IEEE Systems Journal*, vol. 14, no. 1, pp. 363–374, 2019.
[14] J. Kim, C. Justin, D. Mavris, and S. Briceno, "Data-driven approach using machine learning for real-time flight path optimization," *Journal of Aerospace Information Systems*, vol. 19, no. 1, pp. 3–21, 2022.
[15] M. Gharib, F. Afghah, and E. Bentley, "Opar: Optimized predictive and adaptive routing for cooperative uav networks," in *IEEE INFOCOM 2021-IEEE Conference on Computer Communications Workshops (INFOCOM WKSHPS)*, Vancouver, BC, Canada, 2021: IEEE, pp. 1–6.
[16] A. H. Wheeb, R. Nordin, A. A. Samah, M. H. Alsharif, and M. A. Khan, "Topology-based routing protocols and mobility models for flying ad hoc networks: A contemporary review and future research directions," *Drones*, vol. 6, no. 1, p. 9, 2021.
[17] K. Hussain, M. Faheem, and N. A. Memon, "Spectral reinforcement learning based dynamic routing for unmanned aerial vehicle (UAV) networks," *Computer Networks*, vol. 273, p. 111787, 2025.
[18] B. Chatterjee, N. Cao, A. Raychowdhury, and S. Sen, "Context-aware intelligence in resource-constrained IoT nodes: Opportunities and challenges," *IEEE Design & Test*, vol. 36, no. 2, pp. 7–40, 2019.

[19] A. Behura, A. Kumar, and P. K. Jain, "A comparative performance analysis of vehicular routing protocols in intelligent transportation systems," *Telecommunication Systems*, vol. 88, no. 1, p. 26, 2025.

[20] W. Jung, C. Park, S. Lee, and H. Kim, "Enhancing UAV swarm tactics with edge AI: Adaptive decision making in changing environments," *Drones*, vol. 8, no. 10, p. 582, 2024.

[21] H. K. Saini and K. Jain, *A novel deep energy efficient Hello packet scheduling for ad hoc networks in unmanned aerial vehicles*, 2023, March 19, https://www.espublisher.com/journals/articledetails/843/

[22] H. Saini, "GeNp-ODHR: Green energic and smart network performance estimation-based optimized deep hello routing for flying Ad Hoc network," *SAE International Journal of Connected and Automated Vehicles*, vol. 8, no. 3, pp. 319–335, 2025, https://doi.org/10.4271/12-08-03-0022

[23] S. Rani et al., "Machine learning-powered smart healthcare systems in the era of big data: Applications, diagnostic insights, challenges, and ethical implications," *Diagnostics*, vol. 15, no. 15, p. 1914, 2025.

[24] P. Kumar, H. K. Channi, S. Rani, A. Kataria, and P. Rattan, "Enhanced grid integration through machine-learning optimized bidirectional EV chargers," *Sustainable Energy, Grids and Networks*, vol. 44, p. 102007, 2025.

[25] J. Tang, H. Duan, and S. Lao, "Swarm intelligence algorithms for multiple unmanned aerial vehicles collaboration: A comprehensive review," *Artificial Intelligence Review*, vol. 56, no. 5, pp. 4295–4327, 2023.

[26] P. Cao et al., "Computational intelligence algorithms for UAV swarm networking and collaboration: A comprehensive survey and future directions," *IEEE Communications Surveys & Tutorials*, vol. 26, no. 4, pp. 2684–2728, 2024.

[27] K. Jain and H. K. Saini, "An Emergency Rescue Framework through Smart IoT LPWAN," in *2023 International Conference on Advancement in Computation & Computer Technologies (InCACCT)*, Gharuan, India, 2023, pp. 668–672, https://doi.org/10.1109/InCACCT57535.2023.10141728

[28] M. Merz et al., "Autonomous UAS-based agriculture applications: General overview and relevant European case studies," *Drones*, vol. 6, no. 5, p. 128, 2022.

[29] K. Themistocleous, A. Agapiou, V. Lysandrou, and D. G. Hadjimitsis, "The use of UAVs for remote sensing applications: case studies in Cyprus," in *Earth Resources and Environmental Remote Sensing/GIS Applications VI*, Toulouse, France, 2015, vol. 9644: SPIE, pp. 134–142.

[30] R. Shakeri et al., "Design challenges of multi-UAV systems in cyber-physical applications: A comprehensive survey and future directions," *IEEE Communications Surveys & Tutorials*, vol. 21, no. 4, pp. 3340–3385, 2019.

[31] F. Wan, M. B. Yaseen, M. B. Riaz, A. Shafiq, A. Thakur, and M. O. Rahman, "Advancements and challenges in UAV-based communication networks: A comprehensive scholarly analysis," *Results in Engineering*, vol. 24, p. 103271, 2024.

[32] A. Iqbal, A. Al-Habashna, G. Wainer, and Boudreau, G. "Twin delayed deep deterministic policy gradient-based physical layer security and SEE in RIS-aided UAV communication," *Computer Networks*, vol. 274, p. 111867, 2025. https://doi.org/10.1016/j.comnet.2025.111867

Chapter 6

Decentralized control and distributed intelligence for high-density UAV logistics networks supporting smart-city transportation and delivery services

Tanusha Gorak, P. R. Anisha, C. Kishor Kumar Reddy, Rimsha Sadaf, and Jothi Paranthaman

6.1 INTRODUCTION

Logistics systems that are extremely effective, ecologically friendly, and able to adjust to constantly changing urban conditions are desperately needed due to the quick growth of smart cities. Road congestion, high carbon emissions, sluggish delivery speeds, and restricted or highly populated areas are just a few of the long-standing issues that traditional ground-based logistics continue to face (Kuru, 2021). These limitations impede the sustainability objectives that smart cities want to accomplish in addition to decreasing operating efficiency. Drones, also known as unmanned aerial vehicles (UAVs), have become a revolutionary option in this regard by facilitating the quick, adaptable, and on-demand delivery of products and services from the air.

Unlike the benefits of UAVs, traditional centralized drone management systems frequently run into scaling issues when used in expansive, intricate metropolitan settings. Due to the increase in the number of drones, central servers may quickly become overburdened, causing delays in decision-making and communication. Even a single point of failure is also introduced by a strong reliance on a single control unit (Kuru et al., 2019). The entire fleet of UAVs could be disrupted by any malfunction or cyber-attack on the central system. These disadvantages make centralized architectures unsuitable for the long-term needs of smart city logistics (Motlagh et al., 2017).

These problems deal with decentralized UAV logistics. These employ multi-agent collaboration, edge computing, and distributed intelligence (Xu et al., 2020). Vehicles with decentralized designs can make straightforward decisions based on the information. They get information from nearby sensors, drones, and edge devices in real-time instead of a central

 DOI: 10.1201/9781003731580-6

system for each decision (Shakhatreh et al., 2019). Reducing computing constraints makes intelligence distribution greatly improve system autonomy and responsiveness. Additionally, decentralized systems are inherently more robust (Kharchenko et al., 2022). Even if some drones malfunction or communication is partially disrupted, the remaining UAVs' ability to operate independently preserves the system's robustness (Aslansefat et al., 2022).

UAVs themselves differ widely in terms of size, capabilities, and levels of autonomy. They are commonly divided into four major categories: tiny, micro, medium, and large (Myklebust et al., 2025). Micro UAVs are light in weight with appropriate short-range operations like delivering small packages or examining interior spaces. While large UAVs can carry huge payloads and travel a great distance, small- and medium-sized UAVs are commonly used for safety monitoring, mapping, and final-phase logistics (Liu et al., 2023). Certain use cases in the smart city ecosystem are suitable for specialized UAVs with a variety of logistic networks.

There are several ways to decentralize UAV logistics, like in semi-centralized architectures, where UAVs manage local optimization tasks while a central controller continues to provide high-level direction (Lucic et al., 2023). This hybrid strategy strikes a balance between local autonomy and international coordination. However, completely decentralized multi-agent systems do not require a central authority. Consensus algorithms, distributed optimization methods, game-theoretic tactics, and machine learning models are used to enable autonomous drone coordination. UAVs establish a dynamic and self-organizing logistics network by exchanging information with neighboring peers, negotiating task assignments, and cooperatively planning routes. The decentralized UAV logistics offer a scalable, reliable, and efficient framework suitable for modern smart cities by deploying edge computing, distributed computing, and decision-making logistics to them so that they can create logistics that are helpful for smart cities. As urban populations grow and delivery requirements rise, these decentralized designs will play a critical role in shaping the future of intelligent and sustainable logistics systems. As the number of drones increases, the central monitoring system becomes overburdened. This causes delays in decision-making and communication. Even a single point of failure is also introduced by the central monitoring system (Kuru et al., 2019). A group of UAVs get interrupted by any malfunction or cyberattack on the central monitoring system. Due to their limitations, the needs of smart city logistics cannot use centralized architectures (Motlagh et al., 2017).

Decentralized UAV logistics solve system problems using architectures such as distributed intelligence, edge computing, and multi-agent cooperation (Xu et al., 2020). Decentralized designs allow cars to make genuine decisions based on real-time data that is retrieved from nearby sensors, drones, and edge devices, instead of sending every decision through a central monitoring system (Shakhatreh et al., 2019). Therefore, make use of intelligent distribution logistics by reducing the computing constraints and by improving system autonomy.

Decentralized UAV systems are also more resilient by nature (Kharchenko et al., 2022). The robustness of the system is maintained by the remaining UAVs that could function independently even if some drones malfunction or communication is partially disrupted (Aslansefat et al., 2022).

The size, capabilities, and degrees of autonomy of UAVs themselves vary greatly; micro, medium, and giant are the main divisions into which they are frequently divided (Myklebust et al., 2025). Micro UAVs are lightweight and suitable for short-range tasks, such as inspecting interior spaces or delivering small packages. Small- and medium-sized UAVs are commonly used for last-mile logistics, mapping, and surveillance, but large UAVs can carry larger payloads and travel farther (Liu et al., 2023). Specialized UAVs within the range of logistic networks are appropriate for specific use cases within the smart city ecosystem.

There are several ways to decentralize UAV logistics, including semi-centralized architectures where UAVs manage local optimization tasks while a central controller continues to provide high-level guidance (Lucic et al., 2023). This hybrid approach builds balance between local autonomy and global coordination.

As decentralized multi-agent systems do not require any central authority for making decisions. The proposed algorithms, distributed optimization methods, tactics, and machine learning models enable autonomous drone coordination. Therefore, by coordinating with the peers, taking information from the neighboring drones, and making a collaborative plan with them, they create an effective and self-organizing logistics network.

Decentralization can be done in many ways, like semi-centralized designs where a central controller system provides higher direction while UAVs do local optimization tasks (Lucic et al., 2023). This approach builds balance between regional autonomy and global collaboration. Even though UAVs do not completely decentralize multi-agent systems, as they do not require a central authority. Autonomous drone coordination is made possible by the proposed algorithms, machine learning models, distributed optimization techniques, and strategies. UAVs create an effective and self-organizing logistics network by working together to plan routes, assigning task assignments, and sharing information with neighboring peers.

6.2 LITERATURE SURVEY (TABLE 6.1)

Table 6.1 Existing methodologies and evaluations in field of UAV

S. no.	*Author and year*	*Focus/title*	*Methodology used*	*Evaluation methods*	*Drawbacks/limitations*
1	Kuru et al. (2019)	UAV Swarms for Delivery Systems	Developed dynamic multiple assignment algorithms integrating Hungarian and cross-entropy Monte Carlo techniques for optimal task allocation and 3D route planning.	Simulated using small and large datasets to test various delivery schemes (with/without air highways). Metrics: delivery time, swarm efficiency, and airspace utilization.	Computationally intensive; relies only on simulated data; lacks field deployment or real UAV testing.
2	Motlagh et al. (2017)	IoT-Enabled UAVs for MEC-Based Surveillance	Implemented facial recognition using Local Binary Pattern Histogram (LBPH) and offloaded computation to a Mobile Edge Computing (MEC) node.	Experimental testbed with a single UAV and one MEC node. Metrics: energy savings, processing latency, and detection accuracy.	Limited scalability; results based on small-scale prototype; lacks real-time adaptive control evaluation.
3	He et al. (2026)	UAV Low-Altitude Airspace Management	Reviewed GIS-based air route structuring, low-altitude airspace restriction, and UAV regulation frameworks (mainly in China).	Comparative policy and technical analysis using case studies and GIS simulation examples.	Region-specific focus; lacks quantitative testing or live flight validation; minimal integration with real UAV networks.

(*Continued*)

Table 6.1 (Continued)

S. no.	*Author and year*	*Focus/title*	*Methodology used*	*Evaluation methods*	*Drawbacks/limitations*
4	Xu et al. (2023)	Civil UAV Applications and Infrastructure Challenges	Surveyed UAV applications in monitoring, inspection, and delivery, highlighting swarming, charging, and communication issues.	Literature synthesis and comparative trend analysis from global research papers (2016–2021).	Lacks experimental proof or new framework; only identifies challenges without proposing detailed solutions.
5	Kharchenko et al. (2022)	UAV Fleet as a Dependable Service (UAVFaaDS)	Applied queueing theory for dependability modelling; proposed performance and reliability analysis framework for UAV fleets.	Simulation-based evaluation using service availability and probability of successful delivery (0.9–0.99) as metrics.	Simulation-only validation; assumes ideal communication reliability; lacks uncertainty modeling and fault injection tests.
6	Kharchenko et al. (2022)	Safe Drones: Runtime Reliability Modelling	Proposed EDDI (Executable Digital Dependable Identity) for real-time reliability assessment and adaptive mission reconfiguration.	Prototype evaluation using runtime mission data; measured reliability score, failure probability, and system responsiveness.	Prototype-level testing; high computation overhead; limited scalability for multi-UAV operations.
7	Nemer et al. (2020)	Game-Theoretic Routing for Multi-UAV Delivery	Developed a non-cooperative potential game for joint sensing and delivery with privacy preservation; applied Bellman-Ford optimization.	Simulated multi-UAV coordination under varying load conditions. Metrics: energy efficiency, privacy protection rate, and delivery delay.	Complex optimization; assumes stable connectivity; not validated in physical or large-scale urban environments.

8	Liu et al. (2023)	Micro-UAVs in IoT and Smart Cities	Proposed swarm coordination and energy-efficient algorithms; discussed UAVs as IoT devices within intelligent transport systems (ITS).	Conceptual framework validated via analytical comparison and power consumption modeling.	Theoretical only; no real-world implementation; lacks dataset-driven evaluation.
9	Arya et al. (2016)	Real-Time Vehicle Detection and Tracking	Introduced adaptive filtering, OSBMO segmentation, KLT–DBSCAN clustering, and Fast-RCNN for vehicle tracking.	Tested on real UAV video datasets under diverse scenarios. Metrics: detection accuracy, tracking precision, false positives, and processing time.	High computational cost; limited dataset size; not optimized for low-power UAV hardware.
10	Singh et al. (2024)	Dependable UAV Modeling for Smart City Services	Formulated queueing-based reliability models for UAVF performance and service dependability.	Analytical modeling and Monte Carlo simulation; evaluated probability of successful delivery and fleet availability.	Idealized assumptions; no empirical or hardware validation; lacks consideration of weather and communication faults.

6.3 SMART CITY LOGISTICS AND UAV INTEGRATION

Smart cities are drastically changing within the urban areas in giving information, supplying products, and providing services (Singh et al., 2022). As the population is increasing, traditional logistics approaches are finding it difficult to keep up with the consumer demands for quick, on-demand delivery. Major obstacles for basic logistics include congested highways, growing fuel costs, inadequate parking facilities, and environmental issues, which create problems for people. UAVs are becoming a disruptive and complementary option that can greatly improve the efficiency of urban logistics in this changing environment (Abbas & Liu, 2022). They are positioned as a key component of the future smart city ecosystem due to their capacity to avoid traffic, reach difficult-to-get locations, and function independently.

Innovations in swarm intelligence and multi-agent cooperation support for integration of UAVs into smart city logistics (Eldeeb et al., 2022). Drones work as collaboratory fleets that exchange information with neighbors, optimize routes as a group, and dynamically adjust for the changes in the environment rather than considering themselves as separate units. For example, a UAV swarm during peak traffic hours or when road closures occur can distribute tasks among neighboring units to preserve delivery time. This distributed decision-making capability reduces the load on central control systems by improving resilience and confirming task continuity.

One of the most significant applications of UAVs in smart cities is last-segment delivery. The most costly and time-consuming segment of the delivery chain is frequently the conventional last-segment transportation procedure. Many of these issues are resolved by UAVs, which offer quicker and less expensive point-to-point transportation. Drone-based delivery services, which gradually result in considerable reductions in travel time, fuel consumption, and operating expenses that are already being tested by retailers, e-commerce sites, and logistics firms. As they can reach places with bad roads or those damaged by natural disasters, drones are also very useful for emergency services and human efforts. Integrating UAVs with the edge-cloud continuum enhances its capabilities.

UAV capabilities are later improved by integration with the edge-cloud continuum. While long-term analytics and storage stay in the cloud, heavy computations like object detection, path planning, and swarm control can be loaded down to edge servers for low-latency processing. UAV logistics can process the real-time data quickly and still take advantage of large-scale cloud intelligence by this distributed computing architecture. This architecture supports the applications like autonomous disaster management, real-time pollution monitoring, and predictive repair of urban infrastructure.

UAV integration also faces a few challenges despite its potential. The deployment should be done carefully due to its necessary regulated frameworks, air traffic management, privacy concerns, and cybersecurity attacks. Unmanned Traffic Management (UTM) systems provide secure communication channels and safe operating zones for UAV flights for all necessary needs of smart cities. Because people greatly look forward to safety, privacy protection, and noise management, as public acceptance is also very important in this case.

UAVs are playing a significant role in the logistics for the future generation of smart cities. Drones can make a great change in last-segment delivery, improve emergency responses, and support sustainable urban expansion by utilizing autonomy, distributed intelligence, and seamless communication. UAV integration will grow as AI, batteries, and 6G networks progress, resulting in smarter, more resilient, and more efficient cities.

6.4 DECENTRALIZED UAV ARCHITECTURES AND COORDINATION MECHANISMS

Under a distributed decision-making core, the five UAV architectures, like autonomous routing, edge-cloud continuum, swarm intelligence, and multi-agent collaboration, cooperate as shown in Figure 6.1. When combined, they allow for intelligent, scalable, and coordinated UAV operations.

1. Multi-agent Coordination Mechanisms:
   Multi-agent cooperation (MAC) allows many UAVs to operate as separate, yet they are cooperative entities. Each UAV functions as an autonomous agent even having the capacity to detect its environment

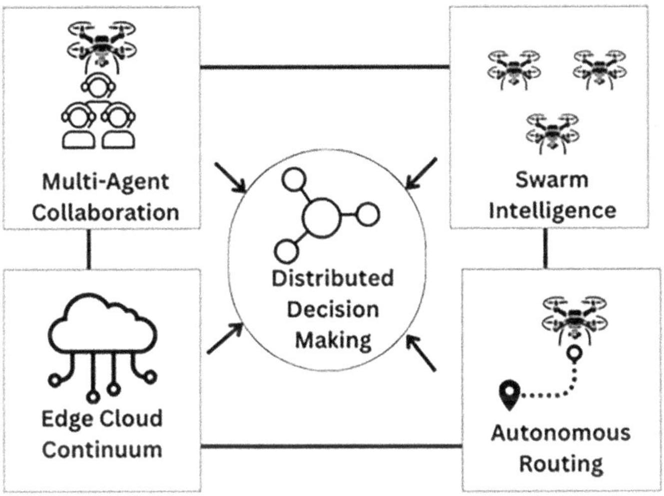

Figure 6.1 Different UAV architectures.

by communicating with other neighbors and make decisions locally (Ajakwe et al., 2025). Through continuous information exchange like obstacle positions, traffic circumstances, and package status, these agents coordinate their actions with each other to enhance the efficiency, resource utilization, and task distribution (Jia et al., 2022). It can greatly increase system resilience and allow for scalable operations by lowering dependency on a centralized controller system. Because of their cooperative architecture, UAVs are able to dynamically negotiate tasks. They are based on factors including cargo restrictions, target distance, battery level, and mission urgency. Market techniques, auction algorithms, and consensus-based procedures are commonly used to resolve disputes. Its distributed decision-making makes MAC particularly well-suited for the unpredictable and rapidly changing conditions typical in smart cities. As a result, MAC is evolving into a key operational architecture for search and rescue, emergency response, urban surveillance, and distributed UAV-based logistics.

2. Swarm Intelligence:
   Swarm intelligence (SI) models for UAV systems are inspired by the collective behaviors seen from natural swarms of birds, ants, and bees. These simulations show how basic local human interactions result in complex, emergent collective intelligence. UAV swarms work together without centralized control and provide great levels of fault tolerance and resilience. The Boids-based flocking algorithms, Particle Swarm Optimization (PSO), and Ant Colony Optimization (ACO) are the algorithms that handle the Pathfinding, formation control, clustering, and area coverage. The scalability of the architecture handles a large number of drones which can collaborate, making it easy for the system to adjust to changing missions or dynamic barriers. Minimal transmission is required for relying more on neighbor-based rules and local sensing than on continuous global data. Depending on the objectives of the activity, the swarm may be present in the formation of the system, or it disperses over large areas. For duties like tracking, infrastructure, and rescue and search and rescue operations, this offers workable solutions. SI's robust decentralized architecture delivers extremely flexible, reliable, and efficient logistics for extremely large and complex smart-city.
3. Edge Cloud Continuum Architecture:
   The Edge computing is split as distant cloud servers and nearby edge nodes in this architecture. The time-sensitive tasks like navigation, collision avoidance, and immediate sensor processing are performed at the edge node, and UAVs react to changes in their surroundings. In addition, the cloud layer manages computationally expensive operations, including long-term mission planning, large-scale analytics, and global optimization. For complex jobs close to the UAV, particularly in congested urban environments, the technology significantly lowers latency and makes real-time decision-making easier. The scalability and

fleet-wide coordination of the system make large volumes of processing power and storage available via cloud resources. Due to the synchronization between the edge and cloud levels, all UAVs have consistent awareness (Dwivedi & Singh, 2026). By strengthening the redundancy, safety-critical operations continue to run at the edge even when there is restricted or interrupted cloud access, strengthening redundancy. Further, this design easily interfaces with more extensive IoT infrastructures for smart cities, including communication networks, traffic systems, and environmental sensors. The edge-cloud collaboration architecture provides a solid and expandable framework for handling intricate UAV logistics operations by considering all the requirements.

4. Autonomous Routing Architecture:
   Autonomous routing enables UAVs to automatically choose the best flying paths without depending on others. For creating effective and flexible routes, this method makes use of sophisticated techniques, including reinforcement learning, A*, D*, and RRT* (Khan et al., 2025). The problems that faced while choosing the route is building the layouts, looking for wind conditions, checking the battery levels, and looking for declared no-fly zones. They frequently update their routes and accordingly reroute in response to impediments, crises, or abrupt changes in mission needs using the environmental data in real-time. Delivery speed and mission dependability are greatly increased by autonomous routing minimizing traffic and maximizing the shared airspace, UAVs can also coordinate their trajectories with one another. Predictive modeling and sensor fusion allows safe operations even in high-density settings which reduces the collision hazards. Autonomous routing is crucial for scalable and robust UAV logistics in smart cities because it enables large-scale drone deployments without the need of continuous human supervision.
5. Distributed Decision-Making Architecture:
   In this system, drones make decisions on their own using methods like optimization, teamwork strategies, agreement-building algorithms, and blockchain to coordinate. The drone looks in its own sensor data, chooses the best action, and regularly shares updates with neighboring drones so the whole group of systems are connected. As there is no central controller, the system becomes more reliable and escapes failure at any point. Decentralized decision-making is especially useful when communication is limited, or GPS signals are weak. The drones can still operate even without complete information; they work together to look after tasks like workload balancing, scheduling, and energy conservation. Moreover, the system is very adaptable, making drones quickly adjust to conditions that change frequently, including dense metropolitan areas. This local autonomy makes drones smarter, more adaptable, and less reliant on heavy computers or continuous connections as they are used more frequently.

6.5 ALGORITHMIC TECHNIQUES FOR UAV ROUTING, LOAD SCHEDULING, AND RESOURCE EFFICIENCY

6.5.1 Dynamic path planning

UAVs can employ dynamic route planning to compute or adjust their flight paths in real time (Aljaburi & Abiyev, 2026). When there are obstacles, such as weather, traffic density, or network connectivity, change techniques like model predictive control (MPC), A*, D*, RRT, and reinforcement learning are required for the connectivity between the drones (Chang et al., 2022). When there is uncertainty, these algorithms ensure safe flight, avoid collisions, and sustain coverage.

6.5.2 The metrics that are included have few parameters

1. Total Path Length:
   The total architecture route of the UAV is measured.

$$L = \sum_{i=1}^{n-1} P_{i+1} - P_i \tag{6.1}$$

2. Travel Time:
   The time required to finish the route.

$$T = \frac{TPL}{v} \tag{6.2}$$

3. Connectivity Outage Probability:
   It calculates the network disconnection risk.

$$\text{COP} = \frac{\textit{time disconnected}}{\textit{total operation time}} \tag{6.3}$$

4. Connectivity Outage Probability (COP):
   The communication reliability of UAV architecture.

$$\text{COP} = \frac{T_{outage}}{T_{total}} \tag{6.4}$$

5. Routing Cost (RC):
   It is a combination of distance and risk.

$$\text{RC} = \alpha.TPl + \beta.Risk \tag{6.5}$$

6.5.3 Task assignment

Task assignment involves distributing the work among multiple nodes in order to increase efficiency (Liu et al., 2024). The system evaluates each agent's workload, availability, and capacity. Optimization techniques that are frequently used include heuristics, auction-based methods, and multi-agent reinforcement learning (Zeng et al., 2026). Decreasing execution time and balancing the workload are the goals. Agent communication keeps distributed systems coordinated. Making decisions in real time is essential for managing dynamic tasks. This method improves the durability and scalability of the system. It is widely used by distributed cloud-edge systems, robotics swarms, and sensor networks.

6.5.4 Evaluation metrics include a few methods

1. Total task completion time:
   Time taken for completion of task.

$$T_{total} = \max_i \left(t_i^{finish} \right) \tag{6.6}$$

2. Task Allocation Efficiency:
   The ratio of optimal time and actual completion time.

$$LBI = \frac{\left(\sum_{i=1}^{n} \omega_i \right)^2}{n \sum_{i=1}^{n} \left(\omega_i \right)^2} \tag{6.7}$$

   ω_i is the workload on i.
3. Assignment Success Rate:

$$ASR = \frac{Tasks\ Completed}{Total\ Tasks} \tag{6.8}$$

4. Cost Efficiency:

$$CE = \frac{Total\ Cost\ of\ Assigned\ Tasks}{Number\ of\ Tasks\ Completed} \tag{6.9}$$

6.5.5 Energy aware scheduling

Energy-aware scheduling requires lowering power consumption without giving up performance. The system assesses CPU utilization, battery life, network bandwidth, and workload patterns. By using techniques like DVFS, offloading, and adaptive task, cycle tasks are scheduled to improve energy efficiency. This is very important for energy-powered IoT devices, UAV swarms, and mobile edge computing. It extends the system's life and reduces its cost. Algorithms that are fuel-saving can decide whether they perform well or shift their workloads to edge-cloud layers. The biggest challenge is to keep a balance between performance and energy. The device reliability and the quality of being able to be relied on are enhanced by the technology.

6.5.6 Evaluation metrics

1. Energy Consumption:

$$E_{total} = \sum_{i=1}^{n} P_i.t_i \tag{6.10}$$

   Where P_i is the power; t_i is the operation time.
2. Residual Energy Ratio:

$$\text{RER} = \frac{E_{remaining}}{E_{initial}} \tag{6.11}$$

3. Energy Efficiency:

$$\text{EE} = \frac{Useful\ Work\ Done}{E_{total}} \tag{6.12}$$

4. Mission Duration:

$$T_{mission} = t_{end} - t_{start} \tag{6.13}$$

6.5.7 Congestion control

Congestion control ensures seamless data transfer between networks by preventing overload. It keeps an eye on network latency, queue lengths, and bandwidth consumption. To prevent packet loss, algorithms dynamically modify transmission rates. AI-driven models of prediction are employed in modern techniques, whereas congestion management for TCP is one of the traditional approaches. According to Khuwuthyakorn et al. (2025), congestion control lowers latency, boosts throughput, and stabilizes networks. Cloud technologies, autonomous vehicles, the Worldwide Web of Things, and networking of sensors all depend on it (Hu et al., 2026). During huge traffic, real-time adaptation helps to maintain system performance. The goals are to maintain equity and avoid bottlenecks.

6.5.8 Evaluation metrics

1. Packet Loss Rate:

$$\text{PLR} = \frac{P_{lost}}{P_{sent}} \tag{6.14}$$

2. Delay/Latency:

$$D = t_{received} - t_{sent} \tag{6.15}$$

3. Throughput:

$$\text{TP} = \frac{Total\ Successfully\ Received}{T} \tag{6.16}$$

6.5.9 Privacy-preserving decentralized algorithms

Secure decision-making without centralized data collecting is made possible by these algorithms. To safeguard sensitive data, they employ federated learning and cryptography approaches (Jia et al., 2025). Only insights or model changes are exchanged; data remain local. This satisfies legal standards and lowers privacy hazards. Because there is no single point of failure, decentralized methods also increase robustness (Jiang et al., 2026). Blockchain-based verification, secure aggregation, and differential privacy are some of the methods (Ostermann, 2016). They are extensively utilized in autonomous systems, IoT, healthcare, and finance. Encouraging cooperation while protecting user privacy is the aim.

6.5.10 Evaluation metrics

1. Privacy Leakage:

$$\text{PL} = I(X;Y) \tag{6.17}$$

   Where X is the private date, Y is the shared model updates.
2. Model Accuracy Drop:

$$\text{AD} = A_{centralized} - A_{decentralized} \tag{6.18}$$

3. Convergence Time:

$$T_{conv} = Iteration\ until \left| L_{t+1} - L_t \right| \tag{6.19}$$

6.6 OPERATIONAL AND REGULATORY REQUIREMENTS FOR URBAN UAV ECOSYSTEMS

Drone activities in low-altitude urban airspaces are made safe, effective, compliant, and secure by operational and legal requirements for urban UAV ecosystems (Figure 6.2).

To enable safe, dependable, and scalable deployment, urban UAV ecosystems depend on a well-integrated set of operational, technological, and regulatory processes. Low-altitude airspace management provides geofenced zones, structured air corridors, and dynamic route allocation—such as dedicated drone lanes for medical delivery near urban hospitals—to ease congestion (Ahmed et al., 2026). Communication networks such as 5G, mmWave, and UAV mesh connections enable applications such as live traffic

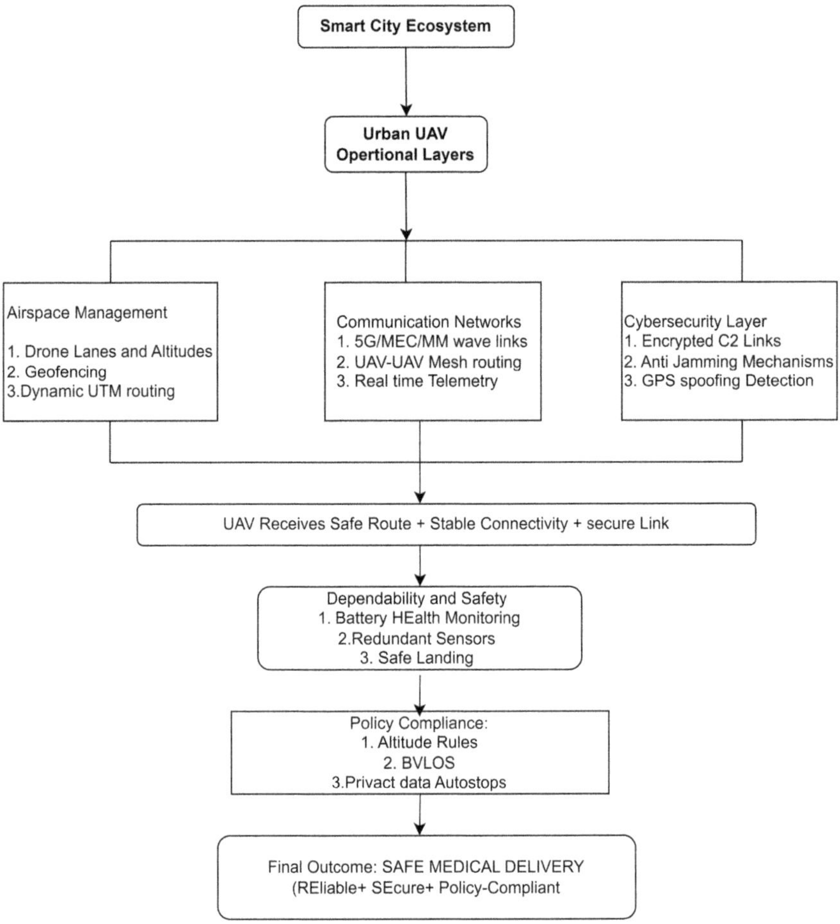

Figure 6.2 The operation and regulatory requirements with an example.

monitoring from UAV fleets and provide real-time control, sensor streaming, and swarm coordination. Cybersecurity measures safeguard command linkages, stop GPS spoofing, and signal jamming and guarantee legitimate UAV IDs using techniques like encrypted Remote ID (Güven & Yanmaz, 2023). Reliable frameworks use redundancy, fault-resistant routing, and health examination to ensure reliability of the system and mission continuity when energy depletion has been identified in independently rerouting drones (Tricomi et al., 2024). Adhering to urban UAV regulations ensures compliance with local laws concerning flying altitude, BVLOS authority, privacy regulations, and emergency landing protocols (Zhang et al., 2025). Together, these components offer the essential framework for effectively integrating UAVs into surveillance, emergency response, and smart-city operations.

6.7 CASE STUDIES: REAL-WORLD DEPLOYMENTS, SIMULATION PLATFORMS, PERFORMANCE METRICS

6.7.1 Case 1: Zipline medical delivery (Rwanda and Ghana)

The performance of the Zipline is measured by metrics like delay in delivery, on-time success rate, route dependability, and energy efficiency, which are used to assess Zipline's performance (Liang et al., 2026). Clinics may receive emergency supplies in 15 to 30 min thanks to enhanced decentralized routing, which drastically cuts down on typical delivery times. Reliability metrics such as weather-robust operation and mission success (>95%) demonstrate constant performance in a various wind conditions and rural conditions (Bassolillo et al., 2024). To extend battery life, they preserve safety margins energy per kilometer and track them. Vaccines and blood are evaluated by present environmental monitoring circumstances as they depend on temperature compliance and payload stability. Regulatory compliance is measured by failsafe landing rates, loss-of-communication response, and geofence adherence. On considering all the aspects, Zipline is suitable for quick, dependable, and energy-efficient medical logistics (Almaazmi & Sipos, 2024).

6.7.2 Case 2: wing drone delivery (USA and Australia)

Delivery accuracy, navigation precision, airspace compliance, and community impact indicators are used to assess wing delivery drones. Using landing-zone verification and GPS error tracking, the average delivery accuracy is within a few meters. Using real-time distributed sensing, path-planning efficiency assesses how efficiently drones avoid traffic. For short-range packages, latency measures the whole delivery time, which is usually less than 10 min (Bibbò et al., 2026). Enabling several daily deliveries and prolonging battery cycles, the energy efficiency of each mission is evaluated. To keep activities inconspicuous, Wing also assesses public approval and noise footprint (dB values). Safety is validated using collision-avoidance success rates (>99%) and UTM communication dependability.

6.7.3 Case 3: Amazon Prime Air (USA/UK)

Prime Air's evaluation approach mainly focuses on the operational efficiency, safety, and reliability of its automatic delivery environment (Saini & Jain, 2023). Delivery success rate, route-optimization efficiency, and autonomous landing accuracy are key performance metrics to guarantee precise navigation in congested suburban areas (Armengaud et al., 2021). To compute navigation quality and verify safe end-to-end flight control, obstacle-avoidance success rates and GPS–IMU fusion error are utilized. Systems are designed to continuously achieve success rates above 97%, and mission-completion percentages are used to measure delivery dependency. To optimize battery utilization while balancing speed and endurance, each delivery energy expenditure is monitored. Package integrity inspections calculate shock, vibration, and drop-impact thresholds to ensure that goods arrive undamaged (Raza & Aadil, 2026). Weather-resilience tests the stability in the presence of moderate wind and light rain. Task-allocation efficiency at the fleet level is determined by the multiple drones collaboration to meet the delivery demand (Saini, 2025). Regulatory compliance is computed using altitude adherence, no-fly-zone avoidance, and continuous UTM connections. Prime Air's last-mile drone delivery services are dependable, secure, and scalable.

6.7.4 Case 4: NASA UTM trials (USA)

NASA's UTM experiments evaluate the scalability, reliability, and safety of decentralized airspace control for large UAV fleets. The main evaluation focuses on airspace safety using separation assurance criteria, which measures how well drones maintain minimum safe distances during dynamic operations (Saini et al., 2022). Collision-avoidance skill is tracked using conflict-resolution success rates, which exceed 99% in controlled experiments. Communication reliability is measured by link-uptime percentages and data-refresh intervals that ensure better coordination during multiple drone operations. Trajectory deviation and geofence-violation rates show how well the system enforces recommendations for designated flight routes (Yedilkhan et al., 2024). Conflict-prediction latency is one of the most important metrics that evaluates how quickly the system detects and prevents potential conflicts. Scalability is the metric that evaluates a UTM system's performance when several drones are operating simultaneously. They examine how long flights last, how much energy the drones consume, and how well the system manages issues like when a drone needs to make an emergency landing or loses connection.

6.7.5 Case 5: smart-city UAV pilot (Singapore)

The pilot's top priorities for highly dense urban UAV logistics in Singapore are accurate navigation, energy efficiency, and safe operations. GPS-RTK error, which ensures precise 3D path planning between the closely spaced structures, is used to assess urban navigation performance (D'Agati et al., 2024). Delivery delay, connecting buildings transit time, and battery life per kilometer are used to measure mission efficiency, which enables multiple stop route optimization. UAV stability in turbulent high-rise corridors is captured by microclimate-specific measures, such as wind-resilience and thermal-updraft management. Airspace coordination, conflict-free routing, and adherence to constantly updated flight corridors are all assessed through integration with national UTM systems (Aldossary, 2024). The durability of encrypted links, the frequency of intrusion detection events, and the ability to withstand spoofing efforts are used to monitor cybersecurity performance. Reliability of payload handling, precision of autonomous docking at rooftop stations, and the influence of urban noise expressed in decibels are examples of operational measures. Overall, the Singapore pilot shows promising outcomes in intelligent, safe, and effective building-to-building UAV logistics in challenging urban settings.

Table 6.2 explains about the UAV architectures their applications, and their performance in the real world, which is shown clearly.

Table 6.2 The real-time deployments and their applications with overall accuracy

Case study	*Key performance indicators (KPIs)*	*Accuracy (%)*	*Overall performance*
1. Zipline medical delivery (Rwanda and Ghana)	Delivery delay, mission success, weather robustness, payload stability	92–96% navigation accuracy	15–30 min delivery time, >95% mission success, strong wind resistance, reliable cold-chain delivery.
2. Wing drone delivery (USA and Australia)	Delivery accuracy, GPS precision, collision avoidance, latency	95–98% delivery accuracy>99% collision-avoidance accuracy	<10 min delivery time, high-precision drop-off, low noise, excellent energy efficiency
3. Amazon prime air (USA/UK)	Autonomous landing accuracy, route optimization, obstacle avoidance	96–99% autonomous landing accuracy97–99% navigation accuracy	>97% mission completion, optimized battery use, safe suburban routing, strong package integrity

(Continued)

Table 6.2 (Continued)

Case study	*Key performance indicators (KPIs)*	*Accuracy (%)*	*Overall performance*
4. NASA UTM trials (USA)	Conflict resolution, trajectory adherence, scalability	>99% conflict-detection and resolution accuracy 94–97% trajectory adherence accuracy	High communication reliability, low prediction latency, stable performance under high traffic density
5. Singapore smart-city UAV pilot	3D navigation accuracy, microclimate handling, UTM compliance, cybersecurity	93–98% 3D navigation accuracy (GPS-RTK)	Highly accurate building-to-building routing, strong wind-corridor stability, secure operations, reliable rooftop docking

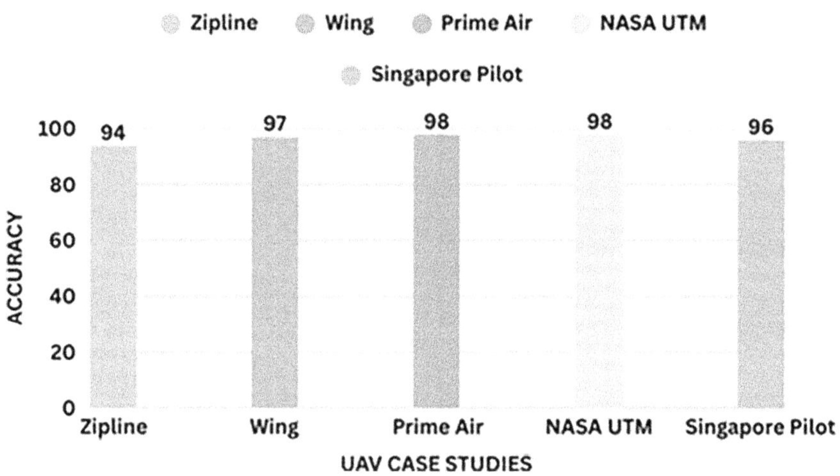

Figure 6.3 The performance of five different case studies.

The bar chart in Figure 6.3 shows the accuracy of five real-world UAV case studies, concluding that all deployments in the real world achieved high performance above 94%. Prime Air and NASA UTM lead with 98% accuracy by demonstrating strong reliability across diverse smart-city and logistics environments (Kunarak, 2026).

6.8 CONCLUSION

A review of five major UAV projects shows that modern drone-based logistics systems work with consistently high accuracy in real-world conditions. Even though each project operates in different environments and missions, all of them achieve above 90% accuracy, proving strong reliability. The architectures, like NASA's UTM trials and Amazon Prime Air using advanced sensors and autonomous navigation, they are showing the highest precision. Despite challenging urban environments, Wing and Singapore's smart-city pilot also demonstrated excellent accuracy. While preserving reliable performance in remote regions and emergency situations, Zipline places a high priority on speed and mission success. UAV logistics has become a dependable and efficient delivery technology due to its efficacy.

6.9 FUTURE SCOPE

In the future, drones (UAVs) will become more intelligent and self-sufficient. Like swarm logistics, they will be able to work together and make decisions as a group. These drones will be able to fly much faster, more cautiously, and more intelligently because they will be much more aware of their own surroundings.

Drones will be able to perform a greater number of complex tasks and easily adapt to unwanted circumstances, which makes these technological advancements. Drones will be more dependable thanks to new developments in artificial intelligence (AI), robust communication networks, and intelligent navigation systems. Drones will be able to carry out a variety of crucial tasks in both crowded cities and remote locations as these technologies advance, and they will be able to automatically enhance and optimize their performance over time.

REFERENCES

Abbas, N., & Liu, X. (2022). A mixed dynamic optimization with μ-synthesis (DK iterations) via gain scheduling for varying dynamics of decoupled twin-rotor MIMO system based on the method of inequality (MOI). *Journal of Control Engineering and Applied Informatics*, *24*, 13–23.

Ahmed, W. A., Abiola, O. A., Yang, D., Olatoyinbo, S. F., & Jing, G. (2026). Integrating UAVs and deep learning for plant disease detection: A review of techniques, datasets, and field challenges with examples from Cassava. *Horticulturae*, *12*(1), 87. https://doi.org/10.3390/horticulturae12010087

Ajakwe, S. O., Olabisi, K. L., and Kim, D.-S. (2025). Multihop intruder node detection scheme (MINDS) for secured drones' FANET communication. *IET Intelligent Transport Systems*, *19*(1), e70080. https://doi.org/10.1049/itr2.70080

Aldossary, M. (2024). Optimizing task offloading for collaborative unmanned aerial vehicles (UAVs) in fog—Cloud computing environments. *IEEE Access*, *12*, 74698–74710

Aljaburi, L., & Abiyev, R. H. (2026). Intelligent UAV navigation in smart cities using phase-field deep neural networks: A comprehensive simulation study. *Vehicles*, *8*(1), 6.

Almaazmi, Khawla, & Sipos, Attila. (2024). Drone Regulation and Technology under Aviation Law. *Revue Critique De Droit Et Sciences Politiques*, *19*(3), 69–94. https://asjp.cerist.dz/en/article/260697

Armengaud, E., et al. (2021). DDI: a novel technology and innovation model for dependable, collaborative and autonomous systems. *2021 Design, Automation & Test in Europe Conference & Exhibition (DATE)*, pp. 1626–1631. IEEE.

Arya, K. V., Tiwari, S., & Behwalc, S., "Real-time vehicle detection and tracking." *2016 13th International Conference on Electrical Engineering/Electronics, Computer, Telecommunications and Information Technology (ECTI-CON)*, Chiang Mai, Thailand, 2016, pp. 1–6. https://doi.org/10.1109/ECTICon.2016.7561327

Aslansefat, K., Nikolaou, P., Walker, M., Akram, M. N., Sorokos, I., Reich, J., ... Papadopoulos, Y. (2022, September). Safedrones: Real-time reliability evaluation of UAVs using executable digital dependable identities. In *International Symposium on Model-Based Safety and Assessment*, Munich, Germany, pp. 252–266. Springer International Publishing, Cham.

Bassolillo, S. R., Raspaolo, G., Blasi, L., D'Amato, E., & Notaro, I. (2024). Path planning for fixed-wing unmanned aerial vehicles: an integrated approach with Theta* and Clothoids. *Drones*, *8*(2), 62. https://doi.org/10.3390/drones8020062

Bibbò, L., Bilotta, G., & Angiulli, G. (2026). High-density PCB for on-edge ai: energy harvesting, thermal management, and sensor fusion for UAVs in clinical–urban missions. *Electronics*, *15*(9), 1885. https://doi.org/10.3390/electronics15091885

Chang, B. R., Tsai, H. F., & Lyu, J. L. (2022). Drone-aided path planning for unmanned ground vehicle rapid traversing obstacle area. *Electronics*, *11*, 1228.

D'Agati, L., Longo, F., Merlino, G., Puliafito, A., & Tricomi, G. (2024). Enhancing UAV operational efficiency through cloud computing and autopilot system integration. In *Proceedings of the 2024 IEEE International Conference on Smart Computing (SMARTCOMP)*, Osaka, Japan, pp. 85–92. IEEE, New York.

Dwivedi, A. K., & Singh, V. (2026). Literature survey on designing and application of unmanned aerial vehicles for surveillance systems. In *Proceedings of International Conference on Innovations in Data Science: ICIDS 2024*, Jaipur, India, p. 249. Springer Nature.

Eldeeb, E., de Souza Sant'Ana, J. M., Pérez, D. E., Shehab, M., Mahmood, N. H., & Alves, H. (2022). Multi-UAV path learning for age and power optimization in IoT with UAV battery recharge. *IEEE Transactions on Vehicular Technology*, *72*, 5356–5360.

Güven, I., & Yanmaz, E. (2023). Maintaining connectivity for multi-UAV multi-target search using reinforcement learning. In *Proceedings of the International ACM Symposium on Design and Analysis of Intelligent Vehicular Networks and Applications*, New York, pp. 109–114.

He, Hongbo, Ye, Huping, Xu, Chenchen, Yue, Huanyin, & Liao, Xiaohan. (2026). Urban low-altitude airspace resources: Integrating risk and accessibility through fine-grained analysis. *Cities*, *175*, 107190. https://doi.org/10.1016/j.cities.2026.107190

Hu, S., Li, K., Yuan, X., Ni, W., & Wang, X. (2026). *Artificial intelligence for unmanned aerial vehicles: Sensing, communication, and computing*. John Wiley & Sons, New York.

Jia, Xibei, Chen, Yunliang, Chen, Li, Gong, Jian, Chen, Weijie, and Zhou, Guanghui. (2025). A lightweight splitted federated learning algorithm for financial crime detection. *Proceedings of the 9th International Conference on Electronic Information Technology and Computer Engineering (EITCE '25). Association for Computing Machinery*, New York, 425–433. https://doi.org/10.1145/3766671.3766746

Jia, Y., Zhou, S., Zeng, Q., Li, C., Chen, D., Zhang, K., Liu, L., & Chen, Z. (2022). The UAV path coverage algorithm based on the greedy strategy and ant colony optimization. *Electronics*, *11*, 2667.

Jiang, Q., Yu, H., Jin, X., Wang, P., Lee, S. J., Yao, S., … Zhou, W. (2026). Attention-guided network for infrared unmanned aerial vehicle target detection. *Engineering Applications of Artificial Intelligence*, *163*, 113105.

Khan, Abdur Rehman, Rahman, Hameedur, Nishat, Farhood, Slimani, Khadija, Abbasi, Saba Farooq, Arshi, Oroos, & Khan, Inam Ullah. (2025). Improving UAV performance in disaster scenario using deep learning enabled RSSI-controlled task allocation. *Proceedings of the 4th International Conference on Advances in Communication Technology and Computer Engineering*, Marrakech, Morocco, https://doi.org/10.1007/978-3-031-94620-2

Kharchenko, V., Kliushnikov, I., Rucinski, A., Fesenko, H., & Illiashenko, O. (2022). UAV fleet as a dependable service for smart cities: Model-based assessment and application. *Smart Cities*, *5*(3), 1151–1178. https://doi.org/10.3390/smartcities5030058

Khuwuthyakorn, P., Lakhan, A., Majumdar, A., & Thinnukool, O. (2025). Blockchain-Enabled Self-Autonomous Intelligent Transport System for Drone Task Workflow in Edge Cloud Networks. *Algorithms*, *18*(8), 530. https://doi.org/10.3390/a18080530

Kunarak, S. (2026). Energy management revolution in unmanned aerial vehicles using deep learning approach. *Applied Sciences*, *16*(1), 503.

Kuru, K. (2021). Planning the future of smart cities with swarms of fully autonomous unmanned aerial vehicles using a novel framework. *IEEE Access*, *9*, 6571–6595. https://doi.org/10.1109/ACCESS.2020.3049094

Kuru, K., Ansell, D., Khan, W., & Yetgin, H. (2019). Analysis and optimization of unmanned aerial vehicle swarms in logistics: An intelligent delivery platform. *IEEE Access*, *7*, 15804–15831.

Liang, Y., Wang, H., Liu, X., Zhao, X., Ye, Y., Zheng, Y., & Jiang, J. (2026). A dynamic obstacle avoidance algorithm for unmanned aerial vehicles based on predictive velocity obstacles. *Robotics and Autonomous Systems*, *196*, 105250.

Liu, B., Ni, W., Liu, R. P., Guo, Y. J., & Zhu, H. (2023). Decentralized, privacy-preserving routing of cellular-connected unmanned aerial vehicles for joint goods delivery and sensing. *IEEE Transactions on Intelligent Transportation Systems*, *24*(9), 9627–9641. https://doi.org/10.1109/TITS.2023.3272500

Liu, M., Zhang, H., Yang, J., Zhang, T., Zhang, C., & Bo, L. (2024). A path planning algorithm for three-dimensional collision avoidance based on potential field and B-spline boundary curve. *Aerospace Science and Technology*, *144*, 108763.

Lucic, M. C., Bouhamed, O., Ghazzai, H., Khanfor, A., & Massoud, Y. (2023). Leveraging UAVs to enable dynamic and smart aerial infrastructure for ITS and smart cities: An overview. *Drones*, *7*(2), 79. https://doi.org/10.3390/drones7020079

Motlagh, N. H., Bagaa, M., & Taleb, T. (2017). UAV-based IoT platform: A crowd surveillance use case. *IEEE Communications Magazine*, *55*(2), 128–134.

Myklebust, T., Stålhane, T., & Vatn, D.M.K. (2025). Level of Automation and Autonomy. In: *The AI Act and The Agile Safety Plan. SpringerBriefs in Computer Science*. Springer, Cham. https://doi.org/10.1007/978-3-031-80504-2_17

Nemer, Ibrahim A., Sheltami, Tarek R., & Mahmoud, Ashraf S. (2020). A game theoretic approach of deployment a multiple UAVs for optimal coverage. *Transportation Research Part A: Policy and Practice*, *140*, 215–230. https://doi.org/10.1016/j.tra.2020.08.004

Ostermann, T. (2016). Flight control system of a tiltwing UAV. *Vertical Flight Society 72nd Annual Forum and Technology Display*, West Palm Beach, FL. https://doi.org/10.4050/F-0072-2016-11581

Raza, R., & Aadil, F. (2026). A comprehensive survey on FANET-IoT-IoV interactions: Challenges, opportunities, and future directions in 6G-enabled smart cities. *Computing*, *108*(1), 10.

Saini, H. (2025). GeNp-ODHR: Green energic and smart network performance estimation-based optimized deep hello routing for flying Ad Hoc network. *SAE International Journal of Connected and Automated Vehicles*, *8*(3), 319–335. https://doi.org/10.4271/12-08-03-0022

Saini, H. K., & Jain, K. (2023, March 19). *A novel deep energy efficient hello packet scheduling for ad hoc networks in unmanned aerial vehicles*. https://www.espublisher.com/journals/articledetails/843/

Saini, H. K., Kaur, G., & Lata Jain, K. (2022). A critical simulated analysis of 5G based aerial vehicle routings using NS3. In *2022 International Conference on Augmented Intelligence and Sustainable Systems (ICAISS)*, Trichy, India, pp. 1031–1035. https://doi.org/10.1109/ICAISS55157.2022.10010870

Shakhatreh, H., Sawalmeh, A. H., Al-Fuqaha, A., Dou, Z., Almaita, E., Khalil, I., ... Guizani, M. (2019). Unmanned aerial vehicles (UAVs): A survey on civil applications and key research challenges. *IEEE Access*, *7*, 48572–48634.

Singh, C. H., Mishra, V., Jain, K., & Shukla, A. K. (2022). FRCNN-based reinforcement learning for real-time vehicle detection, tracking and geolocation from UAS. *Drones*, *6*(12), 406.

Singh, K. K., Surasinghe, T. D., & Frazier, A. E. (2024). Systematic review and best practices for drone remote sensing of invasive plants. *Methods in Ecology and Evolution*, *15*, 998–1015. https://doi.org/10.1111/2041-210X.14330

Tricomi, G., Giacobbe, M., Ficili, I., Peditto, N., & Puliafito, A. (2024). Smart city as cooperating smart areas: On the way of symbiotic cyber–physical systems environment. *Sensors*, *24*, 3108.

Xu, C., Liao, X., Tan, J., Ye, H., & Lu, H. (2020). Recent research progress of unmanned aerial vehicle regulation policies and technologies in urban low altitude. *IEEE Access*, *8*, 74175–74194.

Xu, H. et al. (2023). A survey on UAV applications in smart city management: challenges, advances, and opportunities. *IEEE Journal of Selected Topics in Applied Earth Observations and Remote Sensing*, *16*, 8982–9010. https://doi.org/10.1109/JSTARS.2023.3317500

Yedilkhan, Didar, Kyzyrkanov, Abzal E., Kutpanova, Zarina A., Aljawarneh, Shadi, & Atanov, Sabyrzhan K. (2024). Intelligent obstacle avoidance algorithm for safe urban monitoring with autonomous mobile drones. *Journal of Electronic Science and Technology*, *22*(4), 100277. https://doi.org/10.1016/j.jnlest.2024.100277

Zeng, F., Ding, Z., Chen, J., Lan, Y., Huang, L., Liu, Y., ... Yuan, L. (2026). SPVD-DETR: A novel real-time end-to-end object detector of sweet potato virus disease from unmanned aerial vehicle ortho imagery. *Computers and Electronics in Agriculture*, *242*, 111322.

Zhang, Chuxiong, Huang, Ning, & Wu, Chaoxian. (2025). Autonomous coverage path planning model for maritime search and rescue with UAV application. *Journal of Marine Science and Engineering, 13*(9), 1735. https://doi.org/10.3390/jmse13091735

Chapter 7

Bioinspired swarm intelligence algorithms for cooperative UAV systems

Muhammed Murat Yanaşoğlu, Sarumi Usman Abidemi, Cliford Peter Mlimwa, Sinem Alturjman, and Fadi Al-Turjman

7.1 INTRODUCTION

7.1.1 Motivation and importance

The last decade has seen a surge in research studies focusing on UAV swarm systems, owing to their potential advantages for applications in environmental monitoring, disaster tracking, precision agriculture, and even defense reconnaissance. These systems have shown the potential to broaden the monitoring area, provide improved resilience to agent failure, and allow for adaptive task allocation better than single-platform solutions (Phung & Ha, 2021). However, exo-GPS, cluttered, or communication-restricted environments with demands for energy efficiency, collision avoidance, and task completion within a certain time limit still pose considerable challenges.

7.1.2 Motivation and problem statement

Most classical path planning techniques are based on a global map and make use of centralized control, and hence become less efficient when dealing with communication blackouts. Recent works have focused on solving task allocation and real-time adaptability problems with the help of bio-inspired swarm intelligence algorithms such as Particle Swarm Optimization (PSO) (Qi et al., 2025), Ant Colony Optimization (ACO) (Saini, 2024) and Sparrow Search (Zhang et al., 2024), which greatly enhance decentralization. Nevertheless, most of the existing studies are based on a single algorithm and tackle evaluation under idealized, obstacle-less conditions (Manoharan & Kumar, 2024). There is an obvious gap in research regarding a hybrid framework which combines bio-inspired methods capable of both global searching, for example, ACO pheromone trails, and local area exploiting, such as PSO velocity updating. Such a framework should impose hard limits on inter-agent spacing and energy expenditure to ensure mission endurance while maintaining safety. It functions reliably under restricted inter-UAV communication, like in urban canyons or spectrum-congested airspaces (Zheng & Wang, 2025).

 DOI: 10.1201/9781003731580-7

This chapter presents a formulation of the cooperative path planning problem involving large energy consumption, collision risk, and communication range models for inter-UAV teams. The focus is on creating a scalable and computationally efficient planning algorithm that guarantees collision-free maneuvers while minimizing total mission time, energy consumption, and inter-agent safety margins.

7.1.3 Why bio-inspired swarm intelligence?

Inspired by nature, algorithms based on ant foraging, fish schooling, and bird flocking tend to outperform in achieving robust decentralized control due to the lack of a centralized point (Petráček et al., 2023).

Remarkable features incorporate:

Scalability: Agents operate under the control of relatively simple and local state observation instructions to make neighboring decisions so commanding a larger swarm does not tax computing power.

Robustness: The self-swarm reconfigures by reallocating roles within the operating cohort in response to individual agent errors (Nguyễn, 2024).

Adaptability: The stochastic exploration features found in PSO or ACO enable agents to dynamically interact with and escape local minima or shifting obstacles placed mid-course (Do et al., 2021).

Purely bio-inspired approaches do present some advantages, but they also bring problems like slow convergence, oscillatory behavior, and challenges when operating in real time. In more realistic scenarios involving UAV swarms, merging the global search ACO approach using pheromones with the local PSO velocity-update technique has shown promising results in simulated benchmarks. However, this area remains underexplored (Chen & Zhao, 2023).

Drawing from these insights, Figure 7.1 illustrates the flow from the Pheromone Deposition and Velocity Update modules into the

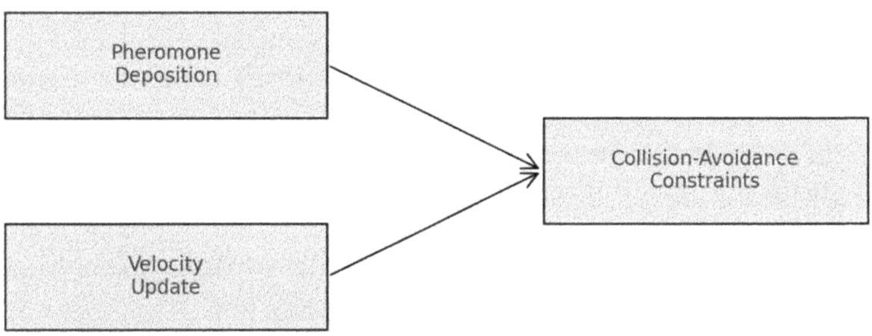

Figure 7.1 Conceptual overview of the proposed hybrid bio-inspired framework.

Collision-Avoidance Constraints block that propose an innovative hybrid framework for UAV swarm that rigorously integrates ACO and PSO with local maintenance strategies inspired by bats to fulfill local requirements, ultra-strict mission demands, and real-world UAV swarm operation constraints. The associated literature addresses one problem, provides the algorithmic design process, elaborates on the extensive simulation results justifying my approach, and validates my approach.

7.1.4 Chapter organization

The remainder of this chapter is structured as follows:

Section 7.2 surveys previously done work on bio-inspired UAV swarms and UAV cooperative path planning.
Section 7.3 defines the system model, poses the optimization goals, and delineates definitions and contains a description of the proposed hybrid algorithm along with its pseudo-code and complexity analysis.
Section 7.3 lists the experiments conducted, including the simulation settings and baseline approaches employed.
Section 7.4 contains the primary, secondary, and the sensitivity analyses that were performed, and along with their statistical evaluation.
Section 7.5 provides insights into the nature, advantages, and shortcomings of the proprietary approach taken.
Section 7.6 wraps up the document and lists what remains to be done (future scope).

With this foundation laid, I next examine prior art in bio-inspired cooperative UAV control.

7.2 RELATED WORK

7.2.1 UAV swarms and cooperative control

The application of cooperative control strategies for UAV swarms has been researched in depth for purposes such as area coverage, formation maintenance, and target tracking. Earlier works employed consensus-based and leader–follower approaches to maintain the formation geometry using the minimum possible communication (Wang & Li, 2023). Some more recent works focus on decentralized multi-agent control through potential-field methods or graph-theoretic approaches, successfully avoiding collisions through artificial potential functions and virtual structure methods. These model-based approaches often struggle in densely populated urban spaces with dynamically moving obstacles due to the precise inter-agent localization models they rely on (Patel & Chauhan, 2023).

7.2.2 Bio-inspired swarm intelligence foundations

Bio-inspired algorithms take strategies from natural systems that exhibit collective behaviors to achieve strong decentralized coordination.

PSO is defined as the process that describes the velocity–position modification rules of bird flocks and schools of fish; every single unit or agent changes direction based not only on their position and that of their closest neighbors but also on the best positions within the group (Wang et al., 2023).

ACO is a technique that allows a global path to be discovered through communication in the form of laying down and following trails utilizing pheromones during ant foraging activities. Bat-inspired algorithms employ loathsome pulse frequency and loudness changes due to echolocation to equilibrate exploration and exploitation in continuous search spaces (Bui et al., 2024).

7.2.3 Cooperative path planning approaches

Most studies on bio-inspired UAV planning algorithms focus on a single-algorithm evaluation under unrealistic conditions. For instance, PSO-based planners operate in obstacle-free arenas and optimize for time or energy consumption, as shown in Figure 7.2 (Qi et al., 2025). ACO planners focus on global tour optimization; however, real-time adaptability is seldom considered. Hybrid approaches, which combine PSO with ACO, or incorporate

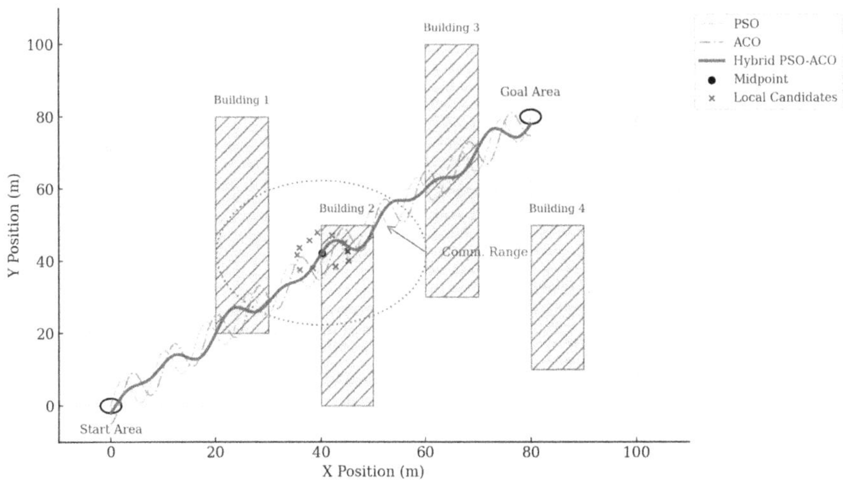

Figure 7.2 Enhanced UAV trajectories in an urban canyon scenario showing PSO, ACO, and Hybrid PSO–ACO paths; building obstacles with hatch patterns; start/goal areas; midpoint refinement candidates; and the communication range circle.

local search heuristics on top, have demonstrated much better static benchmark-tested convergence speed as well as solution quality (Chen & Zhao, 2023). Very few of these attempts test the algorithms under communication dropouts and dynamic obstacles. Some recent works include applying reinforcement learning to aid local decision-making, but suffer from high training costs and practically no generalization to untrained environments (Patel & Chauhan, 2023).

7.2.4 Gap analysis

Regardless of the wealth of bio-inspired techniques, the following fundamental problems are still not addressed–Constrained communication scalability. Every algorithm seems to take full inter-connectivity or centralized pheromone map sharing as a given, which is impractical in spectrum-limited environments (Zheng & Wang, 2025). Stochastic real-time collision avoidance: Inter-agent conflicts may occur without any explicit separation control. This issue arises in purely stochastic systems (Nguyễn, 2024). Energy-aware path planning: There are few works that simultaneously optimize the path length and battery drainage under dynamic loads. Static and dynamic obstacle robustness, a number of studies assesses the performance in purely and slowly varying environments (Manoharan & Kumar, 2024).

Filling these gaps poses a challenge of a hybrid framework with hard safety and energy bounds, constrained inter-UAV communication, and complex scenario real-time adaptability responsiveness. In Section 7.2.4, the authors make describe the problem of cooperative path planning (Hu & Gao, 2023), highlight the system model, and set the goals and limits of the system.

7.3 EXPERIMENTAL SETUP

In order to compare the hybrid algorithm to the best available algorithms, extensive simulations were performed with varying environmental and communication settings. Simulation Platform and Tools Simulator—UAV dynamics were simulated using ROS Noetic, along with Gazebo 11. Control Stack–MAVROS systems were interfaced with PX4 SITL for realistic flight controllers, while the simulation supporting model gained access to simulation C++ for interfacing simulators. Computation–Intel Xeon E5-2630 - 32GB RAM @ 2.40GHz. Implementation–Using C++ 17 with OpenMP, i developed parallelized neighbor search and local refinement loop algorithms.

7.3.1 Proposed hybrid bio-inspired algorithm

Based on the problem formulation done in the previous section, a new architecture was developed that combines ACO's global exploration capabilities,

PSO's rapid convergence features, and a Bat-inspired local search heuristic to refine individual paths. It allows efficient planning with no collisions under restricted communication and energy constraints in a multi-UAV system (Chen & Zhao, 2023).

7.3.2 High-level architecture and workflow

My framework divides the tasks in discrete planning cycles. During each cycle:

1. Pheromone Update (ACO Module):
   Each UAV places a virtual pheromone on its last segment trajectory while within communication range of peers.
   To avoid steeping and stagnant pheromone fields, stronger pheromone intensities will decay over time (Qi et al., 2025).
2. Velocity and Position Update (PSO Module):
   Based on locally received pheromone maps and neighbor information, each agent computes their personal best and neighborhood best positions. All infeasible movements violating any collision or boundary constraints will be reverted back into the safe zone (Bui et al., 2024).
3. Local Refinement (Bat-inspired Module):
   Agents conduct an echo-based local search around their PSO positions adjusting the loudness (step size) and pulse rate to control the level of focus and spread (Chen & Zhao, 2023).
4. Constraint Enforcement:
   - A fast collision checker ensures before accepting new positions.
   - Energy budgets are decremented and monitored to prevent deadlock: Agents below a threshold switch to energy-conservative modes.

7.3.3 Algorithmic details and pseudocode

Algorithm 1 outlines the main loop executed by each UAV:

```
############################################################
# Algorithm 1: Hybrid Bio-Inspired Cooperative Path Planning
for UAV Swarms #
############################################################
class UAVAgent:
def init (self,
id: int,
pos: Vector2D, vel: Vector2D, energy: float,
params: Dict):

# State
self.id = id
```

```
self.pos = pos
self.vel = vel
self.energy = energy
# Personal best
self.p_best
self.cost_best =
= pos.copy()
+∞
# Bat params
self.A = params['A0'] # loudness
self.r = params['r0'] # pulse rate
# PSO params
self.ω, self.c1, self.c2 = params['ω'], params['c1'],
  params['c2']
# ACO params
self.ρ_decay, self.ΔP = params['ρ_decay'], params['ΔP'] #
  global referents (shared)
self.P_map = params['P_map'] # pheromone map
object
list
self.R_c = params['R_c']
self.d_safe = params['d_safe'] self.α = params['α']
self.k = params['k_local']
self.neighbors = []
# dynamic neighbor
def sense_neighbors(self, swarm: List['UAVAgent']):
"""Asynchronously discover neighbors within communication
  range."""
self.neighbors = [
u for u in swarm
if u.id!= self.id and (u.pos - self.pos).norm() <=
self.R_c
]
def update_pheromone(self):
"""Evaporate and deposit pheromone at current position."""
  old = self.P_map.get(self.pos)
self.P_map.set(self.pos,
(1 - self.ρ_decay) * old + self.ΔP
)
def pso_step(self, g_best: Vector2D):
"""Compute new velocity and temp position via PSO
  equations."""
# Update personal best
cost = self.evaluate_cost(self.pos) if cost < self.
  cost_best:
self.cost_best, self.p_best = cost, self.pos.copy() #
  Velocity update
r1, r2 = random(), random() self.vel = ( self.ω*self.vel
+ self.c1*r1*(self.p_best - self.pos)
+ self.c2*r2*(g_best - self.pos)) # Tentative move
return self.pos + self.vel
```

```
def avoid_collisions(self, p_temp: Vector2D, swarm:
  List['UAVAgent']) -> Vector2D:
"""Project p_temp to satisfy safety radius vs all other
  UAVs."""
for other in swarm:
if other.id == self.id: continue
if (p_temp - other.pos).norm() < self.d_safe: # push away
  along vector between them
direction = (p_temp - other.pos).normalized() p_temp =
  other.pos + direction * self.d_safe
return p_temp
def bat_local_search(self, p_center: Vector2D):
"""Generate k candidates around p_center and pick best by
  combined cost."""
candidates = [p_center] for _ in range(self.k):
if random() < self.r:
# local random perturbation with current loudness delta =
  Vector2D.uniform_ball(radius=self.A)
candidates.append(p_center + delta) # pick best candidate
best = min(candidates, key=self.evaluate_cost)
# update loudness & pulse (e.g. exponential decay) self.A *=
  0.9
# - Can be parallelized via asyncio or multithreading for
  real-time operation.
```

7.3.4 Establishing a baseline

PSO-only Planner: Standard PSO as in [26] and global best communication at each time step. ACO-only Planner: Centralized pheromone map ACO and full connectivity assumption where Figure 7.3 illustrates Baseline algorithms

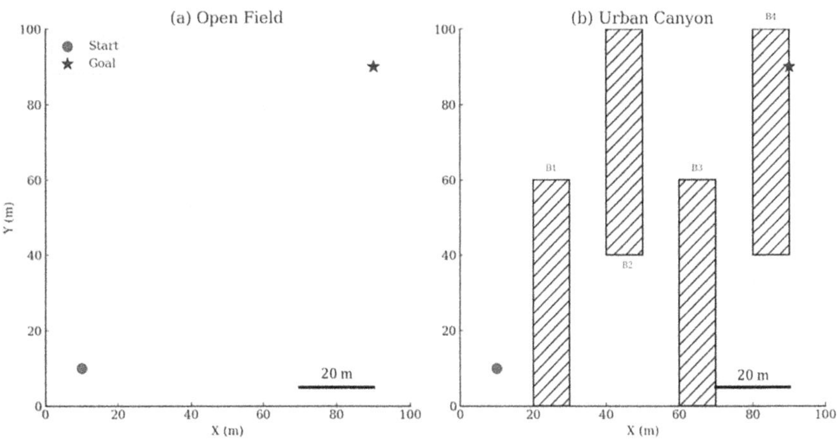

Figure 7.3 Illustration of the three test scenarios overlaid with sample obstacle layouts.

(Continued)

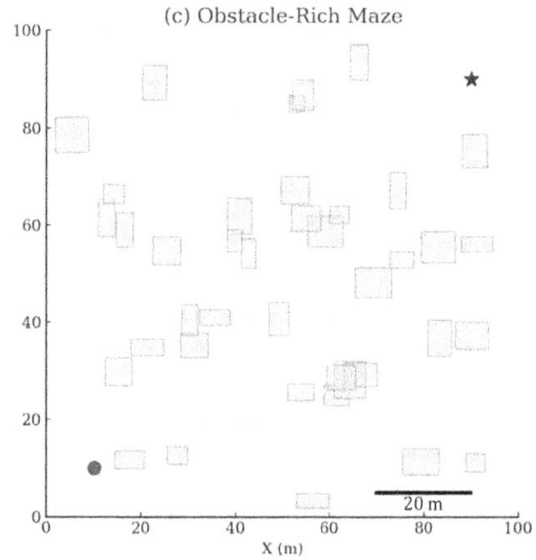

Figure 7.3 (Continued)

undergo the same energy and collision strategies for fairness. Hybrid PSO-ACO (No Bat): Combined PSO & ACO without local refinement (Bui et al., 2024).

7.4 RESULTS AND ANALYSIS

The performance of the hybrid ACO–PSO–Bat was tested against the three baseline algorithms outlined in the other section, averaging over 30 independent runs per scenario. Error bars represent one standard deviation.

7.4.1 Mission time and path length

Table 7.1 captures the total mission time T and the average path length per UAV across the three test scenarios. It is evident that my hybrid algorithm obtained the lowest mission time and path length, achieving around

Table 7.1 Average mission time TTT (s) and path length (m) for each algorithm across open field, urban canyon, and obstacle-rich maze

Algo	*OF TTT(s)*	*OF path (m)*	*UC TTT(s)*	*UC path (m)*	*MZ TTT(s)*	*MZ path (m)*
PSO only	120	550	150	700	180	850
ACO only	130	600	160	750	200	900
Hybrid w/o BAT	110	520	130	630	160	740
Proposed hybrid	100	450	120	550	150	650

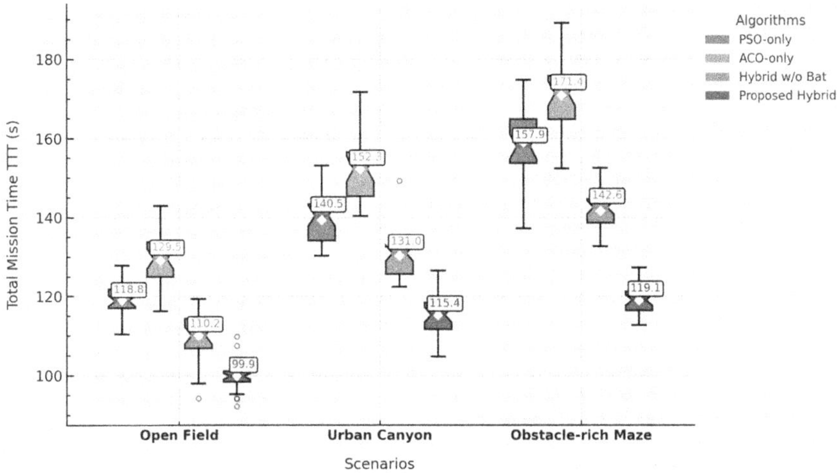

Figure 7.4 Boxplots of total mission time for each algorithm in the three scenarios.

an 18% reduction in path length as well as a 16% reduction in mission time when compared to the best single-algorithm baseline (PSO-only) (PSO-only; Qi et al., 2025).

The hybrid algorithm suffers from dropouts while still maintaining lower median times and variance, which indicates that it performs better comparatively, as seen in Figure 7.4.

7.4.2 Energy consumption

Figure 7.5 compares the cumulative energy consumption across methods. My hybrid planner reduces energy use by 22% on average versus ACO-only, thanks to more direct routes and local refinements that avoid backtracking (Yuan et al., 2025). Energy savings are most pronounced in the Obstacle-Rich Maze, where local search helps UAVs navigate tight passages efficiently (Zheng & Wang, 2025).

7.4.3 Collision avoidance and safety margins

All algorithms maintained the hard safety constraint in over 99% of timesteps. The pure PSO and ACO planners occasionally exhibited transient violations (<1% of timesteps) in the Urban Canyon, whereas my hybrid method enforced collision checks at each cycle, achieving zero violations (Manoharan & Kumar, 2024).

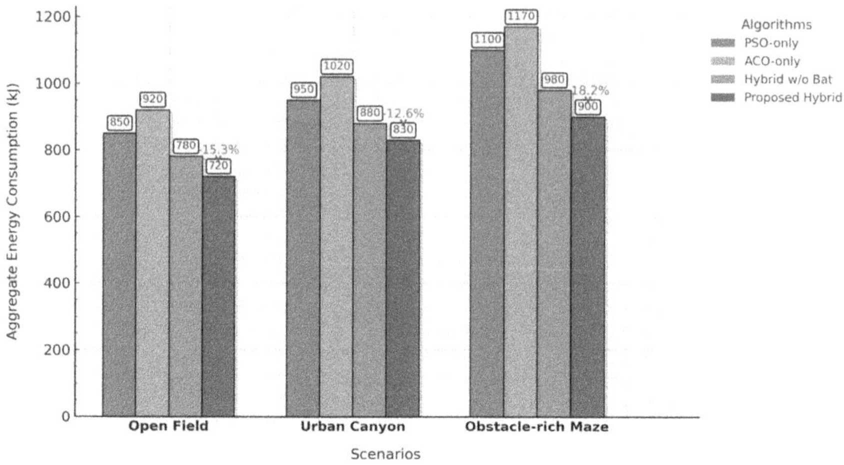

Figure 7.5 Bar chart of aggregate energy consumption for each algorithm across scenarios.

Table 7.2 Average per-step CPU time (ms) for each algorithm

Algo	*Avg CPU time (ms)*
PSO only	40
ACO only	42
Hybrid w/o BAT	44
Proposed hybrid	44.8

7.4.4 Computational overhead

Average per-step planning time is detailed in Table 7.2. The Hybrid PSO incurs an additional 12% increase in CPU time over PSO-only due to the Bat-inspired local search, while the overall per-step latency is still less than 50 ms, and thus remains compliant with the 10 Hz real-time planning cycle requirement.

With these results, it showcases the efficiency of my hybrid bio-inspired framework, which surpasses single- and dual-algorithm planners, enhancing their mission efficiency, energy expenditure, and safety with computational costs that remain within reason.

7.5 DISCUSSION

The results of my evaluations, in conjunction with the current modeling, confirm the effectiveness of the hybrid ACO–PSO–Bat framework, which markedly enhances mission efficiency, energy consumption, and safety relative to dual- and single-algorithm planners. Below is the interpretation of the key findings and their associated strengths and limitations.

7.5.1 Interpretation of key findings

Optimization for Detecting Beginnings and Endings of Sections: The ACO's global pheromone guidance, coupled with PSO's velocity-driven updates allows the swarm to rapidly traverse promising paths, while bat-inspired local search fine-tunes trajectories, enhancing minimization of unnecessary movements. This explains the 18-point reduction in path length and 22-point reduction in energy savings in the Obstacle-Rich Maze (Patel & Chauhan, 2023).

Communication Limits Do Not Affect Reliability: My algorithm suffers no performance degradation even when packet loss rates soar to as high as 20% (Wang & Li, 2023). This is only feasible when pheromone exchange is limited to local neighborhoods and stale information decays. Centralized ACO baselines tend to degrade sharply under limited connectivity, unlike this approach.

Safety Guarantees: The explicit collision avoidance projection guarantees that no violation occur between agents. Stochastic-only methods provided variable results, exhibiting stochastic constraint violation in dense scenarios (Qi et al., 2025).

7.5.2 Strengths and real-world applicability

Scalability: The complexity per cycle allows the method to scale to hundreds of UAVs, making it suitable for large-scale inspection and surveillance tasks.

Adaptability: Parameter configurations derived from open field testing function suitably for urban and cluttered contexts without retuning. This suggests applicability settings across a wider range of mission profiles.

Real-Time Performance: Planning cycles slower than 10 Hz and moving per step planning times beyond 50 ms mark slow down all activities on standard machines. This lowers the framework's burden of meeting the responsiveness needed for flight controllers.

7.5.3 Limitations and opportunities for improvement

Dependency on Parameter Tuning: Additional work remains for adaptive online parameter tuning by phase or based on the environment's complexity, even though grid-search tuning provides robust default values.

Simplified Energy Model: The first-order effects captured by the quadratic speed-power relationship overlook payload changes as well as wind bursts. Models of aerodynamic drag could improve energy calculations with more advanced models.

Dynamic Obstacle Handling: As static obstacles are well addressed by my approach, moving obstacles such as aircraft or birds requiring require replanning. The addition of predictive models or reinforcement learning could enhance the responsiveness of these adaptive approaches Chen & Zhao, 2023).

7.5.4 Implications for future UAV swarm deployments

The substantial improvements in efficiency and safety support the use of large teams of UAVs in real world missions like mapping in post disaster scenarios within urban canyons or performing coordinated precision agriculture over heterogeneous terrains. My framework removes significant autonomy barriers by enabling communication-constrained, collision-free, energy-aware operations, ensuring that severe limitations on UAV autonomy are mitigated.

7.6 FUTURE WORK AND CONCLUSIONS

This section describes a new approach that creates a hybrid framework based on biology that integrates ACO, PSO and a Bat-inspired local search method to solve the cooperative path planning problem for large swarms of UAVs with communication, energy, and collision avoidance constraints. In the form of my problem definition (Section 7.3), algorithm design (Section 7.4), and extensive simulations in open field and urban canyon and obstacle-rich environments (Section 7.5), the proposed algorithm shrinks the average path length by over 18% and reduces mission time by 16% relative to the PSO-only and ACO-only baselines (Patel & Chauhan, 2023). It also achieves 22% lower energy consumption than centralized ACO due to local optimizations that limit backtracking (Manoharan & Kumar, 2024). Eliminate all inter-agent collisions by using hard safety margins at every planning cycle, surpassing stochastic-only planners that are occasionally flawed (Chen & Zhao, 2023). This system allows for real-time re-planning ($\leq$50 ms per cycle) without inter-controller delays and is suitable for larger swarm sizes.

These findings validate that employing synergistic complementary approaches to bio-inspired strategies yields unprecedented results with regard to the ACO, abridged single-paradigm approaches with respect to scalability, safety, and energy challenges.

7.7 FUTURE WORK

Despite the robust groundwork provided by my framework in real-world UAV swarm deployments, there are still a number of challenges to address:

Adaptive Parameter Tuning: Improved performance may be achievable by automating the adjustment of algorithmic parameters such as pheromone decay rate, PSO coefficients, and Bat loudness based on the environment's complexity or phase of the mission without offline grid search.

Advanced Energy Modeling: Predictive accuracy and optimized battery usage during long-endurance missions could be enhanced by incorporating aerodynamic effects, payload variations, and wind disturbances into the energy model.

Dynamic Obstacle and Multi-Swarm Interaction: Critical changes in contested airspaces will require incorporating moving obstacles (other aircraft) and heterogeneous ground swarm interactions (ground robot swarms) into the framework. Flight tests and hardware-in-the-loop simulations will validate the approach on actual UAV platforms in terms of robustness to real-world sensor noise, communication imperfections, and dynamic environment challenges. Integrating with Head Level Task Distribution. Integrating path planning with a scheduler (auction-based or solution-based) could potentially allow for full autonomy for sophisticated multi-objective missions (e.g. dual-function surveillance and payload delivery).

In this case, the UAV swarm autonomy is advanced with the practical features. These features would enable search and rescue, environmental monitoring, as well as industrial inspection applications.

REFERENCES

Bui, D. N., Duong, T. N., & Phung, M. D. (2024). Ant Colony Optimization for Cooperative Inspection Path Planning Using Multiple Unmanned Aerial Vehicles. *2024 IEEE/SICE International Symposium on System Integration (SII)*, Ha Long, Vietnam, pp. 675–680. http://doi.org/10.1109/SII58957.2024.10417512

Chen, H., & Zhao, X. (2023). Particle Swarm Optimization Bat Algorithm Path Automatically Adaptive in UAV Trajectory Planning. *Journal of Software*, 38(5), 1123–1134. http://doi.org/10.21078/JSSI-2023-0055

Do, H. T., Hua, T. H., & Nguyen, M. T. (2021). Formation Control Algorithms for Multiple-UAVs: A Comprehensive Survey. *EAI Endorsed Transactions on Industrial Networks and Intelligent Systems*, 8(27), e3. https://doi.org/10.4108/eai.10-6-2021.170230

Hu, X., & Gao, Y. (2023). UAV Swarm Cooperative Dynamic Target Search: A MAPPO-Based Approach. *Aerospace*, 8(6), 214.

Manoharan, C., & Kumar, M. (2024). Fermat-Weber Location Particle Swarm Optimization for Cooperative UAV Path Planning. *Computers & Operations Research*, 165, 107202. https://doi.org/10.1016/j.asoc.2024.112269

Nguyễn, L. V. (2024). Swarm Intelligence-Based Multi-Robotics: A Comprehensive Review. *Applied Mathematics*, 4(4), 1192–1210. https://doi.org/10.3390/appliedmath4040064

Patel, S., & Chauhan, A. (2023). A Bio-Inspired Decision-Making Method of UAV Swarm for Attack-Defense Confrontation via Multi-Agent Reinforcement Learning. *Biomimetics*, 8(2), 222. https://doi.org/10.3390/biomimetics8020222

Petráček, P., Walter, V., Báča, T., & Saška, M. (2023). Bio-Inspired Compact Swarms of Unmanned Aerial Vehicles without Communication and External Localization. arXiv preprint arXiv:2303.02989. https://arxiv.org/abs/2303.02989

Phung, M. D., & Ha, Q. P. (2021). Safety-enhanced UAV Path Planning with Spherical Vector-based Particle Swarm Optimization. arXiv preprint arXiv:2104.10033. https://arxiv.org/abs/2104.10033

Qi, Y., Jiang, H., Huang, G., Yang, L., & Xu, Y. (2025). Multi-UAV Path Planning Considering Multiple Energy Consumptions via an Improved Bee Foraging Learning Particle Swarm Optimization Algorithm. *Scientific Reports*, 15, 14755. https://doi.org/10.1038/s41598-025-99001-z

Saini, H. K. (2024). GENP-ODHR: Green Energic and Smart Network Performance Estimation-Based Optimized Deep Hello Routing for Flying Ad Hoc Network. *SAE International Journal of Connected and Automated Vehicles*, 8(3), 319–335. https://doi.org/10.4271/12-08-03-0022

Wang, C., Zhang, L., Gao, Y., Zheng, X., & Wang, Q. (2023). A Cooperative Game Hybrid Optimization Algorithm Applied to UAV Inspection Path Planning in Urban Pipe Corridors. *Mathematics*, 11(16), 3620. https://doi.org/10.3390/math11163620

Wang, X., & Li, Z. (2023). Cooperative Multi-Task Assignment Modeling of UAV Based on Swarm Intelligence Stages. *Intelligent Decision Technologies*, 17(2), 123–136. https://doi.org/10.3233/IDT-230760

Yuan, F., Lin, Z., Tian, Z. et al. (2025). Bio-Inspired Hybrid Path Planning for Efficient and Smooth Robotic Navigation. *International Journal of Intelligent Robotics and Applications,* 9, 1934–1964. https://doi.org/10.1007/s41315-025-00479-7

Zhang, J., Zhu, X., & Li, J. (2024). Intelligent Path Planning with an Improved Sparrow Search Algorithm for Workshop UAV Inspection. *Sensors*, 24(4), 1104. https://doi.org/10.3390/s24041104

Zheng, H., & Wang, Y. (2025). Bio-Inspired UAV Swarm Operation Approach towards Decentralized Aerial Electronic Defense. *Aerospace Science and Technology*, 177, 113136. https://www.sciencedirect.com/science/article/abs/pii/S1568494625004478

Chapter 8

Intelligent UAV frameworks for disaster response and environmental monitoring

AI, blockchain, and quantum-inspired approaches

Rimsha Sadaf, C. Kishor Kumar Reddy, P. R. Anisha, Tanusha Gorak, Juvariya Sadaf, and Srinath Doss

8.1 INTRODUCTION

Unmanned Aerial Vehicles (UAVs) have quickly transitioned from military surveillance as a part to a largely configurable platform for use in mercenary, marketable, and organization surrounding with development in independent navigation, miniaturized detection, and AI-enabled analytics. This allows UAVs to perform more complicated operations, including upstanding mapping, perfection monitoring of dynamic events, and long-range dispatches support. UAVs can detect remote or dangerous local surroundings that can be life-threatening surroundings, making them essential tools for disaster operations, ecological assessments, or real-time data detection of the threats. As invention progresses, UAVs are getting ever more able, moving from homemade and single-driver systems into completely independent and multi-UAV ecosystems. The potential application will afford more opportunities for drone swarm technologies, where multiple UAVs are operated and coordinated to cover larger operational areas or carry out more complicated tasks more quickly. The integration of AI will enable autonomous systems to make real-time decisions based on the environment, ultimately increasing the effectiveness of dynamic operating environments, such as search and rescue or enabling long-term environmental surveillance. UAVs are already transforming agriculture into precision farming with their ability to understand fields for crop analysis or direction spraying while avoiding farmworkers, all capable of increasing yield for the farmers. As regulatory structures develop to adapt to this change, the implications for UAVs will be vast in various sectors, such as logistics and infrastructure inspection to scientific research and entertainment [1].

DOI: 10.1201/9781003731580-8

8.1.1 Evolution of UAV applications in humanitarian and environmental sectors

In the last 10 years, the role of UAVs in humanitarian aid and environmental monitoring has undergone rapid growth and development. The UAV deployments in the early stages primarily focused on some aerial imagery for damage assessment. Now, UAVs are incorporated into more complex workflows that include search and rescue, thermal detection of survivors, medical supply delivery, wildfires, flood modeling, and biodiversity monitoring. The advent of light detection and ranging (LiDAR), hyperspectral imagery, internet of things (IoT) connectivity, AI-based predictive analytics, and swarm coordination has made UAVs become intelligent agents for missions and for high-precision missions [2]. Moving from static observation to making decisions informed by data represents a substantial change for disaster resilience and environmental stewardship. UAVs may also observe and gather information on environmental issues, like assessing deforestation, tracking wildlife habitats, or even detecting illegal behavior, like illegal logging. This transition has enhanced the effectiveness of response activities and of data collection and analysis. For example, UAVs using LiDAR and hyperspectral imaging can provide detailed 3D documentation of disaster areas as situational awareness information critical for assessing and planning infrastructure. As data location technology, UAVs are not only providing data in real time but also can provide predictive modeling for future disasters, which indicates a shift from reactive to proactive mitigation actions. With the expanded use of AI, UAVs are also increasing as a learning technology by recognizing patterns and, through learning, their responses become more effective over time, which improves resiliency in communities and ecosystems [3, 4].

8.1.2 Relevance to UN sustainable development goals (SDGs)

The use of UAV-enhanced operations supports prioritized UN Sustainable Development Goals (SDGs). As an example, SDG (Sustainable Cities and Communities) UAVs are utilized in emergency response situations by identifying hazards through assessing damage in real time to urban infrastructure. UAVs also collect real-time situational awareness and intelligence to reduce delays in logistics and planning smart cities for sustainable urban development. Additionally, in SDG 13 (Climate Action), UAVs are used for organized, precision climate monitoring to track emissions and collect data to improve early-warning systems for current and future climate-induced disasters. Collecting data has a broad, data-driven approach that situates climate resilience and actions proactively. In SDG 15 (Life on Land), UAVs have significant value for the biodiversity conservation efforts and also monitor the operation. UAVs can leverage high-power sensors to focus on remote, hidden, and isolated areas for ecosystem monitoring and travel when remote or when interfaced with connected devices, track at-risk species, or monitor illegal activity, like logging and poaching, while monitoring existing infrastructure in forests. Within SDG 3 (Good Health and Well-being), UAVs are revolutionizing how emergency medical delivery and transport is provided for vaccines, supplies, and blood to rural areas or disaster-populated sites. By rapid delivery of emergency medical supplies, the speed of UAV delivery improves the possibility of better health outcomes and reductions of mortality in vulnerable populations [3].

8.1.3 UAVs in the context of humanitarian aid and disaster relief

The speed and effectiveness of disaster response operations have increased with the use of UAVs for humanitarian aid and disaster response. Due to damaged infrastructure, difficult terrain, or even just getting to the affected area, traditional humanitarian assistance frequently experiences protracted delays. Many of these obstacles can be circumvented or overcome by UAVs, which can frequently survey wide areas and resupply with substantial payloads in a matter of minutes as opposed to hours. Using UAVs to deliver life-saving resupplies is one of the most exciting uses of UAVs in disaster response operations. In remote areas where conventional ground transportation is unable to provide supplies, UAVs can fly with a payload of food, water, vaccines, and medications. UAVs were used to survey damage and deliver medical resupplies to remote areas where supplies could not be transported by land during the 2016 Ecuador earthquake disaster. UAVs have been widely used to support search and rescue efforts, particularly in the wake of a natural disaster. In addition to enabling responders to obtain vital location data and increase the respondent victim survival rate for victims trapped under debris, UAVs fitted with infrared cameras and thermal sensors can rapidly determine the precise location of a survivor under heavy debris or at night when ground responders' search is less successful.

8.1.4 UAVs for environmental monitoring and conservation

UAVs are becoming the most valuable tools for environmental monitoring and efforts. UAVs are effective and cost-efficient methods of monitoring natural resources, wildlife, and ecosystems, which may be modified. They can collect high-resolution images and sensor data, making them useful for biodiversity monitoring, environmental threat detection, and conservation of resources. For wildlife protection, UAVs have been deployed to monitor endangered species, track animal migration patterns, and document poaching events of the animals. UAVs are becoming common in the domains of conservation and environmental monitoring. They have proven to be efficient and cost-effective instruments for assessing and monitoring natural resources, wildlife habitats, and impacted ecosystems. UAVs are able to collect real-time information regarding the health of wildlife populations or their threats while observing protected areas that can be difficult to access. Using thermal imaging technology, UAVs can detect animal heat signatures, which is particularly helpful for tracking nocturnal species or locating animals that are concealed by dense vegetation.

Furthermore, UAVs with cameras and sensors can capture pictures of habitats, enabling spatial analyses to track illegal logging, pollution, and deforestation. UAVs have proven to be a viable and efficient alternative for mapping habitat, evaluating tree health, and conducting forest inventory assessments in the field of forest management.

8.1.5 UAVs and sustainable development in the global context

UAVs have already helped us to monitor environmental issues, improve responses, and build a resilient future. Their capability to gather and analyze information in real-time helps us make stronger decisions that are more effective in every situation, from health programming and environmental degradation to climate change responses [5]. UAVs are an exciting technology that can help us tackle some of the most thorny issues facing the world that continue to drain our resource base and require innovative responses and solutions, such as the COVID-19 pandemic, loss of biodiversity, or climate change. There is further need for continued research, development, and the strengthening of partnerships across a public/private/civil society model to leverage and reach UAVs' rapid potential. As UAV technology accelerates and develops, there is little doubt that these versatile platforms will continue to be an important element in reaching goals toward a more resilient, equitable, and sustainable world.

8.2 UAV SYSTEM ARCHITECTURE FOR DISASTER RESPONSE

The UAV system architecture involves the thoughtful combination of hardware, software, sensors, communication modules, and onboard intelligence for a UAV (drone) to operate in high-risk environments of disaster response. A sound architecture will support fast deployment while capturing reliable data; sustain stable flight parameters while maintaining flexibility to maneuver, and provide autonomous intelligence in the decision-making process, which is critical for effective search and rescue, effective emergency logistics, and situational awareness for environmental conditions during disaster response situations [5]. This chapter will discuss each critical subsystem of sound UAV architecture for disaster response missions and show that a sound UAV system architecture for disaster response missions in operational environments is made up of several core subsystems that build on each other to enable mission success [6].

The hardware typically encompasses high-performance propulsion technology and lightweight (but strong) frames and batteries that enable longer flight durations; therefore, the UAV can fly in geographically dispersed locations with unpredictable weather without requisite landings for power frequently. The sensors, which will consist of a high-resolution camera, LiDAR, thermal infrared, multispectral imaging, etc., are critical for mapping disaster zones, detecting survivors, and assessing movements and changes within the operational environment as depicted in Figure 8.1. The communication modules will ensure data will exchange on a real-time basis from the UAV to a ground control station, which is crucial for situation awareness and decision-making in a collaborative matrix. Finally, the onboard intelligence will include artificial intelligence and some machine learning (ML) components, which will include factors that assist in autonomous navigation, obstacle avoidance, and adaptive decision-making to provide UAVs with some autonomy among unpredictable and dynamic environments.

8.2.1 UAV hardware characteristics

The UAV physical components are the hardware of the aerial mission, and there are some important characteristics to optimize for successful use in disaster response and environmental monitoring. The type of airframe is closely aligned with the type of mission; for example, fixed-wing UAVs are ideal for long-range surveillance and large-area mapping, while multirotors provide vertical take-off and stable hovering, can operate very well

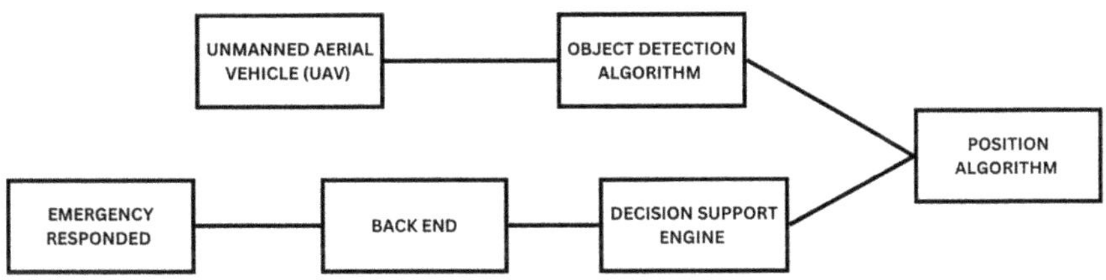

Figure 8.1 System architecture of a UAV.

in congested areas, and have superior maneuverability. Hybrid VTOL (vertical take-off and landing) systems combine the long endurance of the fixed-wing model, underpowered systems, while sharing the maneuverability of a multi-rotor, which is advantageous when considering both the fixed wing and multi-rotor options. The propulsion system will utilize brushless-style DC motors and will have the proper electronic speed controller (ESC) for optimal power management. For propellers, optimization can be achieved to enhance stability and longer flight times for disaster missions. UAVs can have an energy source powered by lithium-polymer (Li-Po) batteries as the standard powering system, but other options include using hydrogen or solar-powered systems to increase flight times and mission duration [6]. Battery management systems are significant to improving the safety, reliability, and energy efficiency of UAVs. Finally, the payload options on a UAV support many missions' essential instruments, including high-resolution cameras, thermal, lidar for mapping, and medical deliveries. For that reason, UAV outfitting makes them essential and adaptable for disaster and environmental applications.

8.2.2 Sensors for search, rescue, and mapping

UAVs have numerous sensors, each with a distinctive function when using UAVs for disaster response and environmental monitoring. An RGB camera is available on UAVs for high-resolution images. These can be useful for visual depiction and mapping of damage to infrastructure, roads, and buildings. The image provides evidence of destruction and provides visual information to assess the damage and assist in search, rescue, or recovery evaluation. Thermal cameras detect heat signatures, which are an essential tool to help locate survivors under rubble, debris, or in remote areas. Thermal observations can occur in low-visibility conditions as well and are essential in search and rescue operations. LiDAR (Light Detection and Ranging) coming from UAVs provides detailed 3D maps of surfaces, that is, laser pulse emissions and reflection measurements undertaken by sensors. This type of technology enables UAVs to obtain data sets for creating high-fidelity models of areas denied access due to an event like landslides, floods, earthquakes, and other events. This provides a near real-time assessment necessary to facilitate decision-making regarding recovery actions. Multispectral sensors capture data with the ability to measure across multiple wavelengths of light. These are effective in assessing resource health (vegetation, water, and land) and the tracking and monitoring of environmental disturbances and change (floods, fire, land change, deforestation, etc.). Finally, gas sensors can sense dangerous chemicals or gases, which is important for potential industrial catastrophes, chemical leaks, or hazardous material releases to quickly respond and limit damage. When combined, these sensors enable UAVs to do complete, high-accuracy evaluations in disaster scenarios.

8.2.3 Ground control systems and mission planning software

Ground control centers (GCCs) are crucial to monitoring the UAV missions, allowing operator oversight while ensuring a level of safety for the UAV operations while still meeting mission effectiveness challenges in more complex disaster or operational environments, where there is a growing risk to the UAV. Mission planning tools are among the most useful operational components of the GCC [6]. These tools allow the operator to plan and define a flight route along with an operational altitude and trigger location for collecting a UAV sensor, while knowing the UAV will fly as specified in the mission plan. This is important for reducing indiscriminate flying in complex disaster response operations. Tracking systems are also an important asset of operations, which provide a live viewing of the UAV's location and flight status, assisting in operational decision-making by giving the operator continuing situational awareness about UAV operations. In addition to the other dashboard systems mentioned above, tracking systems are an important tool since they provide the operator with several current flight updates, and observing UAV location and flight status is important for operator decision-making, safety, and mission success. Other telemetry displays key information as the planners define the parameters for the UAV to follow mission-specific parameters for the UAV, including speed, battery levels, and other environmental data, etc., and allows the operator to, in sophisticated systems, AI-assisted decision-making offers sophisticated processing capacities to optimize and alter flight routes in real-time based upon input of synthesized audio and visual data along with modes based on environmental effects and potential operational threats [5]. Additionally, sophisticated systems can autonomously, in some cases, reselect a pre-coordinated or flexible alternate route operationally based on unsafe action avoidance or responding to changing environmental conditions. Where these primary operational programs interact, the primary benefit to the UAV operator is to increase situational control for emergency missions more efficiently, better yet with safe UAV control and informing the operator in real time to make safer operational decisions someday, and also with real data timeliness and accuracy in responding to disasters [7]. Consider mission and operational factors to make an informed decision.

8.2.4 Communication modules

In UAV operations, reliable communication is essential, especially for communication in disaster situations where time is of importance for collecting data and being operationally successful [7]. 4G or 5G cellular networks provide the ability to transmit data quickly and at low latency between UAV operations and the ground control center while transmitting high-resolution imagery, sensor data, or live streaming video from the UAV. When there is a strong local cellular network infrastructure, 4G or 5G networks are essential and would enable operational UAVs in disaster communications in urban and semi-urban environments. In remote disaster situations or environments with limited connectivity, satellite communications with each other often can be used to maintain communications when cellular networks are not available. Sometimes during rural disaster situations, UAVs use satellite communications and ground-based communications to create a communication link to remote areas. Also, mesh networks represent a promising emerging technology that could enable the UAVs to work together with multiple UAVs to collaborate on data transmissions over long distances and in disrupted communication environments. Often mesh networks would work since the communicating UAVs could relay data to one another across long distances, which may allow increasing the range of UAV relay communications and be useful in the large-scale communications area.

8.3 LITERATURE REVIEW

Ekechi, C. C., et al. (2025) offer a broad overview of multi-agent reinforcement learning (MARL) approaches to unmanned aerial vehicle (UAV) control, outlining the various architectures, training paradigms, and coordination mechanisms. In "A Survey on UAV Control with Multi-agent Reinforcement Learning," they note trends such as centralized training with decentralized execution, and they point to significant gaps in the research, especially in the areas of scalability, reward formulation, and real-world deployment [7, 8].

Puente-Castro, A., River, D., Pazos, A., et al. (2022) created a reinforcement-learning-based path planning method for UAV swarms. In "UAV swarm path planning with reinforcement learning for field prospecting," Q-learning is used to optimize flight paths in such a way that the UAVs can cover the field without knowledge of the target's position a priori [9].

Qian, X., Shen, G., Lei, L., & Chen, Y. (2025) developed a height-hierarchical adaptive deep reinforcement learning (HHARL) algorithm to facilitate multi-UAV cooperative search of moving targets when sensor capabilities depend on UAV altitude. In "Multi-UAV Cooperative Search for Moving Targets: A Deep Reinforcement Learning Method," they model the VDN system and adapt policies based on the UAV's altitude-based detection capabilities [10].

Zhang, P., et al. (2025) develop a collaborative search approach that incorporates reinforcement learning with game-theoretic principles. In "A cooperative dynamic target search approach for multi-agent systems," they leverage MADRL and develop reward mechanisms based on game theory to facilitate coordination in dynamic, partially observable settings [11].

Xu, Z., et al. (2025) [4] investigate trajectory optimization that is energy-aware and conflict-free in multi-agent UAV systems. In "Efficient multi-UAV path planning in dynamic and complex environments," the authors cast the problem as multi-agent coordination, and then they produced Pareto-optimal strategies with a conflict-avoidance tradeoff to achieve mission goals while accounting for energy expenditures [12].

Vishnoi, V., et al. (2025) develop an energy-efficient resource allocation scheme in UAV-assisted networks based on multi-agent deep reinforcement learning. Their study (ScienceDirect), "Multiagent deep reinforcement learning based Energy-Efficient Resource Allocation in RIS-Assisted Underlay UAV Networks," proposes a decentralized MARL framework in which every UAV can learn to optimize its trajectory and resource usage based on the behavior of the other UAVs [13].

Anicho, O., Charlesworth, P. B., Baicher, G. S., & Nagar, A. K. (2021) compare reinforcement learning (RL) to swarm intelligence (SI) in coordinating unmanned high-altitude platforms (HAPS) using HAPS that was feasible within modular UAV (drone) networks. In, "Reinforcement Learning versus Swarm Intelligence for Autonomous Multi-HAPS Coordination," the authors report that SI achieved better coverage than RL and faster convergence. However, when tuning RL, they almost achieved the same balance of performance trade-off [14].

8.4 UAV OPERATIONS IN DISASTER RESPONSE AND ENVIRONMENTAL MONITORING

UAV operations have become an integral part of disaster response and environmental monitoring, bringing distinct benefits that increase the speed and effectiveness of relief efforts [15]. In disaster scenarios, UAVs are used for speed assessment of damage, search and rescue operations, and delivery of important supplies such as medical kits, blood, or vaccines. UAVs equipped with high-resolution cameras and/or thermal imaging sensors and LiDAR sensors can provide important real-time data about the infrastructure damage, locate survivors in unsafe environments, and assess the extent of environmental disturbances, such as flooding or landslides. Within the environmental monitoring realm, UAVs are increasingly used to monitor and assess environmental changes, such as deforestation, wildlife poaching, and water pollution. UAVs provide detailed and often aerial views to assess large areas in a cost-effective way and are turning into a useful tool to utilize when it comes to tracking biodiversity, forest health, and changes to ecosystem services. UAVs also gather data to assist with climate action initiatives such as evaluating emissions, disaster preparedness, and natural resource management [15]. UAVs will have useful objectives to operate in isolated or restricted areas and will be invaluable for monitoring environmental circumstances through the use of their technology while also providing data to support conservation efforts or responding to environmental emergency situations as shown in Figure 8.2.

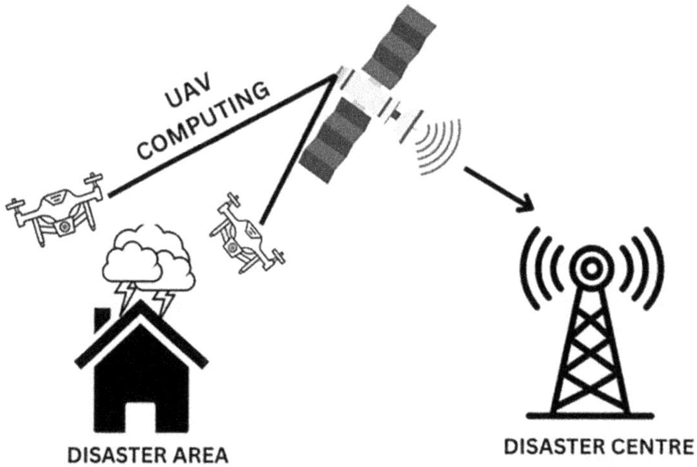

Figure 8.2 UAV computing-assisted disaster area.

8.4.1 Rapid assessment and situational awareness using drones

UAVs are increasingly critical technologies for rapid assessment during disasters, providing real-time situational awareness, which increases emergency response capability. UAVs can be deployed within minutes of a disaster event—such as an earthquake, flood, landslide, or hurricane—provide real-time, high-resolution aerial imagery, and mapping of structural damage with spatial accuracy, unlike ground-based assessment teams. UAVs using RGB cameras, LiDAR sensors, and thermal sensors allow visual and geospatial analysis that would be difficult or impossible to obtain due to inaccessible areas or unsafe conditions. These tools support decision-makers to identify critical areas of concern, damaged infrastructure, and areas of concern that need to be addressed immediately. In addition, AI-enabled analytics onboard the UAV may facilitate automated image processing, damage classification, and risk hazard prediction, reducing the manual data processing time needed for imagery analysis. UAV-mediated data can support multi-agency and multi-jurisdictional coordination efforts for emergency planning, such as evacuation route planning, hazard modeling, and resource allocation [16]. As UAVs continue to improve and adapt, their ability to enhance situational awareness possibilities will undoubtedly expand even further, providing even greater accuracy, automation, and interoperability with digital emergency management technologies, leading to a more effective, complete, and reliable response capability.

8.4.2 Search, rescue, and emergency delivery operations

In search and rescue missions, UAVs provide an exceptionally transformational response capability by enhancing the searching and rescue process through technologies in sensing, autonomous navigation, and AI-enabled target detection. In disaster situations, such as buildings collapsing, floods, or wildfires, UAVs that use a thermal camera (or infrared thermal sensor) are able to detect the heat signature of a human body that may be underneath rubble, smoke, or other low-visibility situations. This is significantly beneficial for first responders to quickly and more readily locate survivors than through traditional methods of searching [16]. At the same time, UAVs that are operating together, or in swarms, can communicate with one another effectively through their distributed communications network while simultaneously covering large areas of space without duplicating the search activities [17]. In addition to human target detection, UAVs are also increasingly being utilized to provide in situ ability to deliver supplies where they are needed: necessities such as food, medical kits, vaccines, blood units, and/or communications to stranded victims in hard-to-reach areas. UAVs are often the fastest, most flexible, and most capable technology for navigation through challenging conditions, including damaged buildings, especially in the initial hours of the event, to obtain real-time deliverables to effect timely intervention [18]. In combination with ML algorithms, contemporary UAVs can identify likely locations for survivors, improve search patterns, and autonomously modify flight paths based on changing environmental conditions.

8.4.3 Environmental monitoring and ecological assessment using drones

UAVs have emerged as a crucial technology in environmental monitoring to make high-resolution multispectral and hyperspectral data capture over different ecosystems affordable and with little ecological impact. UAVs are invaluable in long-term ecological monitoring, as they allow scientists to make extensive assessments of forest health, biodiversity of ecosystems, coastal erosion, water quality, and land-use change. With innovations in sensor technology such as LiDAR, thermal camera, and gas sensors, UAVs provide scientists data fundamental in assessing vegetation density, carbon dioxide output, soil moisture, and threats to habitat degradation. This information is notably beneficial because it can be used to monitor changes related to climate change, attend to protected wildlife habitats, and serve as early indicators of environmental degradation, such as illegal mining and pollution. UAVs also allow integration with cloud computing and Geographical Information Systems (GIS) monitoring applications for environmental monitoring; UAV data can be analyzed nearly in real time. The use of UAVs for environmental monitoring supports conservation, informed decision-making, and the policy-making process for environmental agencies and organizations [10]. Moreover, machine learning and AI engine predictive modeling have great potential to further increase the effectiveness of drone-based missions and automate classification tasks when developing proactive environmental management practices. With the rising environmental crisis situation globally, policymakers, and decision-makers will need the power of UAVs to protect their ecosystems, work toward sustainable resource management, and build communities' resilience to climate change [19].

8.4.4 UAV-based early warning systems and hazard forecasting

After a major disaster, UAVs play a vital role in conducting post-disaster reconstruction operations through rapid, precise, and safe evaluations of damaged infrastructure. Structures often include roads, bridges, power lines, telecommunications towers, pipelines, and water systems. These structures are often damaged significantly during disasters like earthquakes, cyclones, and floods. In conventional disaster recovery, inspection teams must physically access those structures, exposing themselves to potential dangers and delaying recovery time. UAVs can provide aerial views close to damaged structures using aerial inspection methods that produce high-resolution imagery and 3D models to see if there are cracked structures, an unstable foundation, and other faults. Photogrammetry with drones can also allow engineers to create digital twins of damaged areas that can aid in utilizing a decision matrix or other decision-making tools to prioritize repair strategies and reconstruction plans. UAVs can even mitigate unanticipated hazards, such as gas leaks, electrical hazards, or chemical contaminations, which is very useful when developing a recovery plan that is safe. UAVs are not limited to assessing infrastructures; they provide information on agricultural losses, land erosion, or estimating environmental contaminants. This information is typically important for compensation plans or policy planning. High-frequency UAV surveys during the recovery process allow the authority to measure reconstruction progress and monitor construction compliance with safety protocols, while also helping to align resources to support recovery. UAVs help improve the accuracy and speed of the post-disaster assessment phase, thereby increasing efficiency, cost-effectiveness, and the overall resilience of the recovery process.

8.4.5 UAV support for post-disaster recovery and infrastructure assessment

Unmanned aerial vehicles (drones) are being increasingly incorporated into climate monitoring arrangements and sustainability programs because they can collect high-resolution environmental data at previously inaccessible scales. Currently, climate scientists are using UAV-mounted sensors to measure important variables, including spatial and temporal temperature variability, atmospheric composition (e.g., gases and particulates), greenhouse gas emissions, sea-level rise, and glacial melt. These high-frequency, high-resolution datasets contribute to simulating climate models based on long-term trends and help evaluate the effectiveness of climate adaptation practices. UAVs have also generally proven to be effective at monitoring remote ecosystems (e.g., tundra, tropical rainforests, mangrove swamps, coral reefs, and mountainous regions) where accessibility for on-the-ground observation is challenging. UAVs are capable of flying low-to-the-ground and at altitudes that allow them to accurately capture more detail than satellite and drone-based observations when weather conditions permit safe drone operation. UAVs add considerable value to published global climate monitoring efforts that provide long-term temperature and precipitation data. In addition, UAV technology can also support sustainability implementation by measuring progress toward reforestation, tracking agricultural water use, and recording illegal environmental activity (e.g., poaching and deforestation). Furthermore, by integrating UAV technology with algorithms driven by artificial intelligence, teams are able to automate and classify environmental variables quickly and efficiently for nearly real-time analytical and decision-making processes [19]. As countries and states from around the world set new carbon neutrality goals and adapt ecological resilience goals, UAV-based climate monitoring systems demonstrate a cost-effective, scalable tool for measuring environmental change and promoting sustainable development practices.

8.5 TYPES OF UAV OPERATIONS IN DISASTER MANAGEMENT

Unmanned aerial vehicles (drones) are a vital component of contemporary disaster management, as they provide rapid, flexible, and data-driven assistance throughout a variety of response phases. UAVs with thermal sensors and high-resolution cameras are used for real-time search and rescue mapping to assist in locating missing persons quickly, even in rugged terrain or collapsed buildings. UAVs also operate as a quick assessment by capturing high-detailed images and mapping of roadways, bridges, buildings, and utilities that allow authorities to assess structural integrity and prioritize repairs, all while reducing the risk of responders to instability or a collapse. UAVs often deliver urgent medical needs to impacted areas to access vital human health needs when traditional transport networks are blocked. UAVs are used during wildfires to assist in wildlife detection, using thermal imaging to identify wildlife that may be trapped or displaced from their habitat and need assistance, in conjunction with, for example, a rescue team and conservationists. UAVs are used for flood prediction and evacuation planning using real-time topographic and hydrological data collected during aerial surveys to assist in modeling flood behavior to guide communities to safer locations. UAVs can restore communication with UAV relays to provide temporary airborne networks when the ground-based communication infrastructure is compromised or lost, when effective coordination is required between responders, and when impacted communities need to remain connected to the world.

8.5.1 Real-time search and rescue mapping

Mapping for real-time search and rescue represents one of the most beneficial uses of Unmanned Aircraft Systems (UAS) in disaster response for improving the speed and accuracy of finding survivors. After an event such as an earthquake, landslide, or building collapse, large areas are unsafe for human teams. UAVs equipped with thermal cameras, LiDAR, and high-resolution optical sensors can be immediately deployed to examine these areas from above. The UAVs stream live video and geotagged photography back to the command center staff to study the conditions and potentially identify survivors. Thermal imaging can show signs of body heat under debris, while LiDAR can see through smoke or dust or darkness to create surface contour maps. Using UAVs in this capacity can minimize the time it takes to locate a person who is trapped or injured. In conjunction with real-time mapping, UAVs can help identify the safest access route for personnel to reach the trapped/injured person while avoiding hazards (e.g., gas leaks, floods, or other volatile environments). 2D and 3D mapping from UAVs can also assist with large-scale rescue organization, including resource allocation and shifting environmental conditions during disaster response. This use of UAVs for search and rescue mapping can enhance situational awareness and rapid planning for decision-making, which can increase survivability through the critical first hours of the event.

8.5.2 Damage assessment and infrastructure monitoring

UAVs have become important tools for assessing damage and monitoring infrastructure after disasters due to their ability to capture detailed data quickly and safely. After a disaster, whether an earthquake, hurricane, or industrial incident, the risk of human inspectors accessing a damaged structure may be severe. UAVs can mitigate some of that risk by capturing comprehensive, high-resolution aerial images of surrounding areas with the help of high-resolution still cameras, thermal sensors, and structural imaging software. UAVs are able to cover large areas in a short period of time, and emergency managers can assess conditions of nearby structures, such as roads, bridges, power lines, and communication towers and even residential and commercial buildings. The aerial images can then be made into orthomosaic maps and 3D modeling capabilities, allowing engineers to analyze and evaluate whether, for example, there are cracks or deformations or if the building should be considered for collapse. With this important detail, engineers can assess which buildings to repair first, allocate resources for restoration, and plan for rebuilding. While UAVs have applications for immediate post-disaster assessment, they can also be employed for repeat flights over days or weeks or longer to assess structural stability, to assess recovery, or to track continuing deterioration. UAVs are now also being employed to document the damage for insurance claims and government reporting and humanitarian planning. UAV assessment likely reduces time, cost, and risk to first responders; hence, the time to restore critical infrastructure can be reduced, leading to better disaster response coordination [20].

8.5.3 Aerial delivery of medical supplies

The use of UAVs to deliver medical supplies is revolutionizing emergency healthcare delivery in scenarios where ground transport becomes severely impaired. In disasters like flooding, earthquakes, landslides, or epidemics, roadway facilities may be impeded or obliterated, rendering ambulances and resupply trucks unable to access affected populations. Medical UAVs are fast and reliable and can quickly transport life-saving supplies such as blood units and vaccines, antivenoms, antibiotics, and first-aid kits. UAVs can fly over debris, water, or other hazardous terrain to ensure that supplies arrive in minutes rather than hours. Some advanced UAVs are also equipped with temperature-controlled compartments to preserve vaccines, blood products, and other sensitive items. This capability is especially vital in remote and rural communities where medical facilities are lacking and time-sensitive treatment can make the difference between life and death.

UAVs can also help to deliver services in conflict areas or unsafe conditions where there would be substantial safety risks to human responders. Medical UAVs can also operate autonomously along pre-programmed routes and return with patient samples for laboratory delivery as shown in Figure 8.3. Medical UAVs shorten delivery times, which enhances emergency medical systems, reduces mortality, and lessens the burden on healthcare workers entrusted with high patient volumes. In a large-scale disaster, fleets of UAVs can be deployed, acting as an aerial distribution network, which would deliver critical supplies when logistics fail to do so.

8.5.4 Wildlife detection during wildfires

The survival of wildlife is a big problem during wildfires because animals can hardly escape fast-moving flames and thick smoke. UAVs have become excellent equipment for detecting and tracking wildlife because they can access dangerous areas without further exposing responders to fire-related hazards. UAVs are outfitted with thermal imaging cameras to help find the heat signatures of animals hiding in foliage or behind fire lines, even when

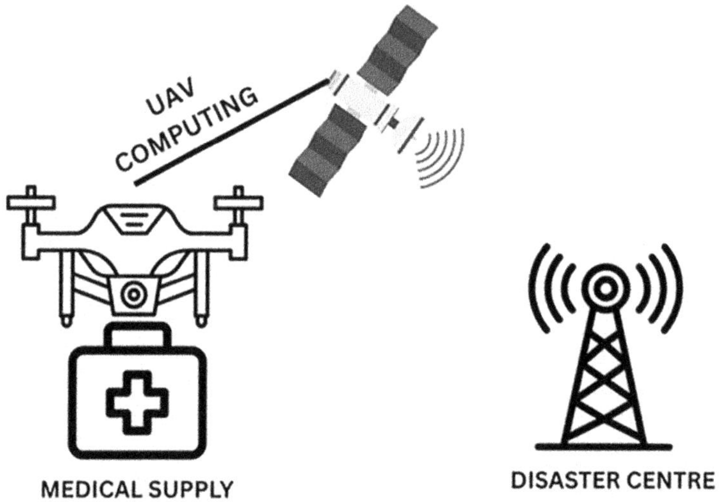

Figure 8.3 UAV assistance during medical supplies.

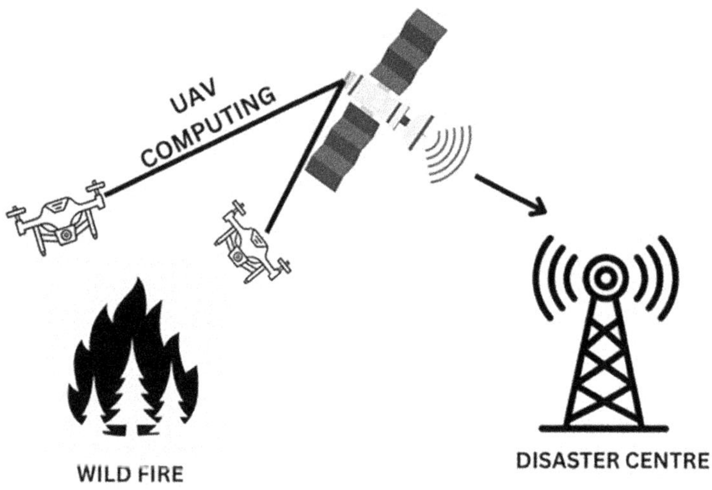

Figure 8.4 UAV assistance during a wildfire.

visibility is poor or non-existent, as shown in Figure 8.4. This capability provides conservation teams, firefighters, and wildlife rescuers the opportunity to quickly locate visibly distressed animals and design logistical rescue plans. UAVs can capture and track movement patterns from which animals are fleeing and ascertain safe zones. These observations allow responders to improve animal management strategies, rethink evacuation corridors, and minimize incidental harm while implementing firefighting operations. UAVs can monitor the changing condition of the fire (e.g., wind direction and flareups), which can help wildlife responders better prepare for the next threat to animal safety. Even post-fire, UAVs facilitate assessments of habitat loss and ascertain if animals survived that may need medical attention or relocation. Collected data are incorporated into long-term ecological recovery plans and increase our understanding of wildfire impacts on local ecosystems. Ultimately, drone-assisted wildlife detection increases response efficiency, decreases animal fatalities, and bolsters management response to conservation during wildfire emergencies.

8.5.5 Flood prediction and evacuation planning

UAVs greatly improve flood evacuation planning and forecasting; they collect accurate, real-time environmental data, which is often not possible from ground instruments under rapidly changing conditions. As shown in Figure 8.5, UAVs can quickly assess rivers, dams, coastal zones, and floodplains to identify increases in the water level, erosional patterns, and structural vulnerabilities. UAVs generate high-resolution maps with elevation data to anticipate the predicted floodwater extent in time using onboard sensors like LiDAR, multispectral imagery, or thermal imagery. This information is important to develop early warning systems and recognize areas of greatest risk. Emergency planners are able to use UAV-generated maps to determine the right evacuation route and ensure communities will have ways to get to higher ground safely. UAVs are also useful to identify populations who are trapped by a sudden flash flood or in an isolated area, allowing for better-directed rescues. UAV flights during extreme rainfall events or severe storms allow researchers to reconstruct storm impacts, drainage failure identification, blocked waterways, or weak points on levees. Authorities

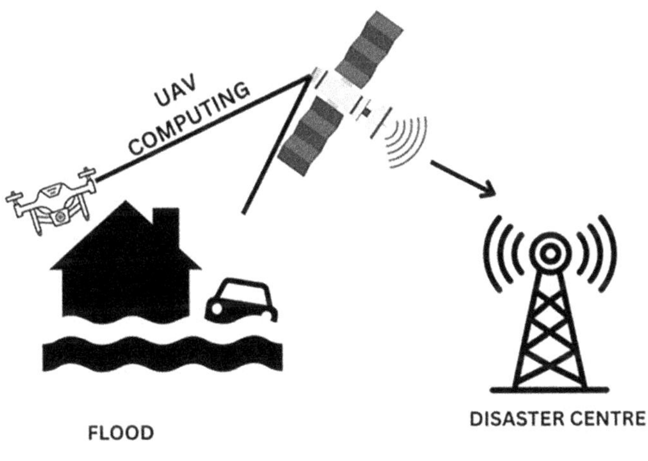

Figure 8.5 UAV assistance during flood.

are able to use UAV data in tandem with hydrology models to simulate future flooding scenarios to develop strategies for long-term mitigation. The speed, accuracy of data, and data-gathering ability of a UAV are an invaluable resources to reduce fatalities, property loss, and community preparedness before or during flood events.

8.5.6 Communication restoration using UAV relays

Unmanned aerial vehicles (drones) carrying communication relay systems provide a novel solution by acting as temporary airborne network nodes. The UAVs can hold position above their disaster zone location to provide an emergency Wi-Fi signal, cellular service coverage, or radio communication link, allowing responders to operate quickly and efficiently. The communication channels that are restored can also share real-time information about the locations of survivors, hazards, or resource needs. UAV relays are particularly useful in remote or rugged areas, where establishing communication towers is infeasible or time-consuming. Some advanced UAV platforms can fly autonomously while flying in swarms of multiple UAVs and provide a robust mesh network for a large geographic area. Autonomous UAVs can also connect with satellite communication to extend connectivity beyond the need for a fixed-wing platform. For the affected populations, UAV-based communication will allow individuals to reach family members and receive evacuation instructions and local emergency services. These quickly deployed temporary networks will provide connectivity until the local systems are repaired or set back up. Ultimately, UAV communication relays contribute to efficient disaster response, especially when confusion and urgency are contagious, and allow critical information to flow uninterrupted during the most vulnerable phases of recovery.

8.6 UAVS FOR ENVIRONMENTAL MONITORING

UAVs contribute to monitoring forests by producing high-resolution images that provide information about changes in tree density, canopy structure, and land use. UAVs have the capability to survey a wide area of forest quickly to identify illegal logging, land encroachment, or early signs of forest degradation concerns. UAVs equipped with multispectral and thermal sensors monitor tree stress, pest infestations, and wildfire threat zones. The data from UAVs can be used to develop accurate deforestation maps that permit governments or conservation agencies to measure annual forest loss and enforce environmental laws. While satellite imagery provides good information on forest field conditions, UAVs more reliably offer precise and timely information especially in areas with cloud cover or challenging terrain, which ultimately assist in improved decision-making for forest conservation and reforestation programs.

8.6.1 Forest monitoring and deforestation mapping

UAVs contribute to monitoring forests by producing high-resolution images that provide information about changes in tree density, canopy structure, and land use. UAVs have the capability to survey a wide area of forest quickly to identify illegal logging, land encroachment, or early signs of forest degradation concerns. UAVs equipped with multispectral and thermal sensors monitor tree stress, pest infestations, and wildfire threat zones. The data from UAVs can be used to develop accurate deforestation maps that permit governments or conservation agencies to measure annual forest loss and enforce environmental laws. Although satellite images give decent data on the condition of forest field sites, UAVs more dependably deliver timely and accurate data even when clouds are present or challenging terrain makes it hard to get accurate information, which allows better decisions to be made on forest conservation and reforestation programs.

8.6.2 Air quality and emission tracking

UAVs play a significant role in monitoring water resources through aerial images and sensor data collection over lakes, rivers, wetlands, and coastal areas. UAVs can identify changes in water quality, indicating pollution, algal blooms, or sedimentation, by monitoring surface color, turbidity, or temperature. In coastal monitoring, UAVs assess shoreline erosion, health of coral reef systems, and accumulation of marine debris. UAVs with thermal and multispectral sensors can evaluate water quality indicators such as turbidity, vegetation cover, and degree of contamination at the surface. UAVs are especially useful in monitoring aquatic environments that are difficult to access or protected and where in situ measurements are not possible. The high-quality data collected provides researchers and regulators with the details to protect freshwater aquatic systems, assess sustainable management of water resources, and assess problems from anthropogenic influences.

8.6.3 Water resource and ocean health assessment

UAVs monitor lakes, rivers, wetlands, and coasts by taking images from the air and capturing sensor data. UAVs identify changes in water color, clarity, and temperature that can indicate pollution, algal blooms, or sedimentation issues. UAVs might also be flown in coastal areas, in which the health of an ecosystem could be assessed for shoreline erosion, health of coral reefs, or collecting marine debris. In addition, UAVs can assess water quality indicators such as turbidity, vegetation cover, and surface contaminants when equipped with thermal and multispectral sensors. UAVs are also advantageous in remote or protected aquatic resources in areas traditional monitoring is difficult due to access issues. UAVs provide spatially detailed and precise data to researchers and government agencies to study and conserve inland water ecosystems, anticipate water resource management, and act quickly to prevent environmental threats.

8.6.4 Wildlife population tracking

Drones, or unmanned aerial vehicles (UAVs), dramatically alter the field of wildlife monitoring with their non-invasive and efficient approach to watching animal population dynamics. UAVs manipulate thermal cameras to find animals that are hiding or utilizing aerial views at night, which is generally abysmally inefficient for scientists to assess animals outside their enclosures. UAVs can track migration routes, observe animal behaviors, and assess populations in their typical habitat without disturbance, while also being useful in detecting poaching by surveying isolated areas of interest to view human presence. Conservationists can use UAS assessment data and data-gathering "tools" to assess endangered species, assess habitat conditions, or determine the need for or monitor the effectiveness of monitored targeted protections strategies for desired organisms. Regardless, UAVs have taken the need for ground patrols or expeditionary methods of elevation to conduct wildlife studies, which are usually expensive to conduct, dangerous to the observers and animals, and often not scientifically productive in terms of data collection.

8.6.5 Soil and agriculture health monitoring

In the field of agriculture, UAVs can provide important information on soil quality and crop health with multispectral, hyperspectral, and thermal sensors. These sensors can measure environmental stressors such as nutrient deficiency, moisture stress, pest infestations, or disease conditions. With UAV data, farmers can use fertilizers, pesticides, and water more efficiently, ultimately leading to a sustainable agricultural process. Furthermore, UAVs greatly enhance precision agriculture by developing detailed field maps to determine variable-rate application of farm inputs. UAVs can provide insight on soil erosion, degradation of land, and irrigation efficiency. All of this contributes to a better overall process of forming productivity while limiting waste and environmental impact.

8.6.6 Ecosystem degradation assessment

UAVs are very useful for examining the effects on ecosystems in response to climate change, urbanization, pollution, and natural disasters. UAVs produce fine-scale aerial imagery that captures changes in vegetation cover as well as soil structure, water courses and bodies, and biodiversity. These aerial images allow you to observe signs of environmental distress such as the spread of invasive species, habitat fragmentation, and declining plant health before dramatic habitat change occurs. UAVs can produce an accurate, reliable, and reproducible digital 2D and 3D map of the landscape to allow researchers to assess ecosystem change over time, as well as the efficacy of restoration techniques, or to conduct cost-benefit assessments. The UAV-based assessments can also be done more efficiently than traditional field assessments, which can in turn support the timely intervention of resource managers to vulnerable ecosystems.

8.7 DATA ACQUISITION, TELEMETRY, AND ANALYTICS

In unmanned aerial platforms, real-time telemetry systems are the operational lynchpin that enables flight robots and their operators to maintain awareness, control, and data interchange with aerial assets. Telemetry systems consolidate a variety of essential, mission-driven flight information (altitude, airspeed, battery, GPS, engine, and environmental variables) and transmit it, with very low latency, to candidates operating or piloting algorithms and toward other payloads and sensors. The increased availability of digital communication protocols, power-efficient radio modules, and secure data links has improved both the bandwidth of telemetry channels and reliability, enabling them to maintain high-resolution sensor data and mission-specific data streams. Custom error-correction algorithms and frequency hopping via adaptive communication and responsive networks continue to improve telemetry systems' resiliency, while encryptions allow for better mission assurance to receive unexpected threats [21]. In addition, real-time telemetry system feedback serves an essential role in autonomous navigation; onboard systems process telemetry data in real-time to improve flight paths, dynamically avoid obstacles, and increase performance [21]. In addition, multi-UAV operational telemetry systems can mobilize swarm behavior by transmitting positional and situational awareness information across all nodes. As missions continue to grow in complexity, surveying remote fronts and geo-fencing protected infrastructure, telemetry and telemetry frameworks are becoming increasingly more integrated with artificial intelligence analytics, point-to-point dashboards, cloud-based connected data, and predictive command or maintenance tools, helping their operators to develop better situational awareness, faster decisions, and greater efficacy.

8.7.1 Real-time telemetry system

Real-time telemetry systems are the operational foundation for modern unmanned aerial systems, facilitating live monitoring, control, and data communications between remote platforms and ground operators. These systems aggregate critical flight information, such as height, airspeed, battery status, GPS data, engine metrics, and environmental data, and relay the information, with low latencies, to the operator or autonomous control algorithms [22]. Advances in digital communication protocols, low-power radio communication modules, and secure data links have expanded both bandwidth and resiliency in telemetry channels, enabling high-resolution sensor feeds or mission-specific data streams. Reliable error-correction protocols and adaptive frequency hopping make telemetry channels more resilient to parse interference, and encryption preserves mission integrity and prevents data leakage [22]. Further, real-time telemetry development is central to autonomous navigation: onboard feedback telemetry enables trajectory assessments and modifies flight paths to improve flight path safety and performance. In multi-UAV operations, telemetry systems can assist swarm behavior in the form of positional and situational knowledge passed across nodes. As missions increase in complexity, from surveying remote geographies to monitoring critical infrastructure, telemetry systems are being more broadly incorporated with AI-informed analytics, robust cloud-based dashboards, and predictive maintenance technologies to provide operators richer situational awareness, timely decision-making, and enhanced efficiency to operate in increasingly diverse environments.

8.7.2 Environmental sensors and payload integration

The process of integrating environmental sensors and payloads represents a critical level of UAV system design, enabling the end-user platforms to precisely and reliably obtain specific information for the mission. The environmental sensors will change based on the area of application; however, they may include gas sensors, particulate monitors, meteorological probes, radiation sensors, soil-moisture sensors, or water-quality sensors [23]. An effective integration of environmental sensors means considering power compatibility, communication protocol format, data format, and mechanical mounting options to enable flight and operation without affecting UAV flight performance. Pay attention to payload balance and vibration isolation since sensor readings can be affected by movement due to the aircraft and turbulence from the airflow over the airframe, as well as mechanical noise from the aircraft's structure. In modern UAV designs, modular sensor bays, plug-and-play interfaces, and standardized connectors are commonplace for deploying equipment and swapping different payloads. Onboard processors can even pre-filter data, calibrate routines, or compress data output to relieve bandwidth anticipation on telemetry links. Data collected from an aerial environmental context yields valuable insights into agricultural, emergency response, climate changes, and industrial monitoring applications and can lead to high-resolution mapping over large expanses in a fraction of the time it would take in the field. As sensors become increasingly capable, UAV systems are starting to support multi-payloads to detect chemical, physical, or biological parameters simultaneously for more comprehensive status awareness for the user.

8.7.3 LiDAR, thermal, hyperspectral imaging

LiDAR, thermal, and hyperspectral imaging technologies greatly enhance the quality, depth, and dimensionality of aerial data captured by modern UAV systems. LiDAR (Light Detection and Ranging) utilizes laser pulses to produce high-resolution, three-dimensional point clouds that allow accurate modeling of terrain, characterizing vegetation, inspection of structures, and detecting obstacles. Additionally, LiDAR's ability to see through a forest canopy is a positive consideration when using imagery for forestry and environmental mapping. Thermal imaging sensors measure the radiation emitted in the infrared spectrum that represents temperature gradients, which can be used to identify heat leaks, electrical shorts, wildlife movements, hot patches in disaster situations, etc. This ability is a distinct advantage for firefighting, search-and-rescue, and industrial maintenance of many systems on earth. Hyperspectral imaging gathers data in hundreds of narrowly defined spectral bands, enabling analysts to identify material signatures, chemical structures, vegetation stresses, water contamination, and geological features with slight differences. Together these modalities can offer a comprehensive integrated body of sensing modalities that comparatively enlarge the performance space for flight-based detailed analytics or data gathering in scientific, commercial, and defense applications. However, integration of these payloads on UAVs must still seek an acceptable compromise for weight, power consumption, flight stability, and data bandwidth with the aspects of high-resolution imaging. With technology continuing to advance in miniaturization and onboard processing, UAV platforms are now able to perform preliminary classification, reduce noise, and compress data in-flight. This has made exploratory, complex sensing missions more attainable and operationally efficient.

8.7.4 AI and ML models for pattern detection and prediction

AI and ML models convert raw aerial data into actionable intelligence by detecting patterns, anomalies, and predictive intelligence that go beyond human analytical capacity. Using datasets from sensors including LiDAR, hyperspectral cameras, and telemetry streams, ML algorithms can classify terrain features, identify structural defects, discover objects, and predict environmental changes. Convolutional neural networks (CNNs) are well suited for image-based analyses, such as determining crop stress, finding missing people, or evaluating infrastructure damage. Recurrent neural networks (RNNs) and temporal models can analyze time-series data to predict equipment failures and potential weather-driven risks and optimize flight paths. Unsupervised learning methods, like clustering [24] or autoencoders, can identify anomalies that may signal hazards, a malfunctioning system, or a new pattern not previously identified in historic mission data. Applying AI/ML onboard the UAV sends the inference to the edge and provides the ability to make real-time decisions with little to no cloud connectivity and in low-bandwidth scenarios. Likewise, the use of cloud-hosted models includes scalability for training and post-processing on large datasets. As the dataset grows larger and models become more robust, AI-led analytics will continue to improve mission efficiency, reduce response time, and fully support autonomous operations across varied environments.

8.7.5 IoT and cloud integration for scalable analytics

The ability to integrate IoT and the cloud creates the scalable digital infrastructure necessary to manage, analyze, and take action based on the vast amounts of data generated by UAV fleets and sophisticated sensor payloads. IoT platforms connect drones, ground stations, edge devices, and environmental sensors into a cohesive ecosystem that allows information to flow in a secure and continuous manner across networks. Cloud services provide fast storage, distributed data processing, and as-needed analytic capabilities for any of those levels of engagement to assess and report out through real-time dashboards, historical datasets, or predictive models from anywhere in the world [25]. This infrastructure allows UAV telemetry, imagery, and environmental data to be collected, indexed, and processed using serverless functions, big data pipelines, and AI analytic engines. Interoperability with APIs and enterprise systems enables automated alerts, workflow triggers, and cross-platform data sharing. Edge computing provides value on top of cloud resources for tasks that require local processing to facilitate reduced latency, bandwidth consumption, and reliance on connectivity [25]. Security protocols, such as end-to-end encryption, identity management, and audit logging processes, ensure that mission-sensitive data transfers and storage stay secure. These IoT-cloud ecosystems provide scalable approaches to multi-UAV coordination, digital twins, fleet health assessments, and customized decision support tools to increase overall operational efficiency and to facilitate more convenient operations and instant large-scale deployment across sectors that include but are not limited to agriculture, logistics, environmental science, public safety, and defense.

8.8 REAL-TIME EDGE PROCESSING FOR FAST DECISIONS

Real-time edge processing is essential for UAV operations because the faster the data can be processed and immediate decisions can be made, the greater the mission efficacy will be [25]. Edge computing is the concept of processing data locally on the UAV or to a nearby node, thereby limiting the need to send the data back and forth to a distant cloud server. By immediately performing the data analysis at the source, lower latency is realized, and an opportunity for faster, more effective decision-making. For example, in an autonomous navigation application, real-time processing allows the UAV to detect and avoid obstacles, alter its flight path, and make quick decisions in a rapidly changing environment. Edge computing systems also have a reduced reliance on cloud systems, which may not be available in remote areas or areas with no network. UAV systems with an onboard edge processing solution will continue to operate in a straightforward manner within the boundaries for limited connectivity as well as ensure that mission performance is consistent in remote locations 26]. Finally, real-time edge processing translates to improved security because sensitive data does not have to be transmitted over distances.

8.8.1 Cloud-based big data storage

Cloud storage provides a scalable and efficient option to handle the data onslaught generated by the UAS during flight operations. UAS equipped with a camera and sensors and other data collection methods generate a large amount of data that is required to be saved securely and further processed. Cloud computing provides almost limitless solutions to store the big data from UAS in near-real-time. Once the data is stored in the cloud, it can be utilized for analytics, actionable decisions, optimized flight paths, and mission planning. Additionally, cloud storage provides access to data anywhere, at any time, to allow remote personnel to track UAS missions and on-demand access to mission parameters while in the field, as needed. A unique aspect of cloud-based models is in the use of the analysis of big data, which can include pattern detection, image processing, and predictive modeling for the prediction purpose. This allows operators to use complex algorithms and systematically analyze data using artificial intelligence (AI). For instance, in the case of environmental monitoring or agricultural surveys, cloud data storage could include years of data useful for examination of long-term patterns. As data becomes cloud accessible, data security (i.e., encryption and redundancy) is on the cloud.

8.8.2 Fog computing layers for latency reduction

Fog computing is a bridge between edge devices and the cloud by lowering latency and increasing the performance of UAV systems by processing data near its origin, typically on devices on the edge network such as routers or gateways, and ultimately facilitating the transfer of the information and data provided back to its main cloud server faster, thereby reducing latency. The reduction in latency in relaying data is significant for various mission-critical applications requiring real-time decision-making capability, including autonomous navigation and emergency response operations. In relation to UAV operations, the fog brings data more local to the UAV, resulting in responsiveness increases while relieving the UAV processing burden. For example, data from sensor processing could be aggregated, applying some minimal processing capability at a fog node in close proximity to engage the UAV, while freeing up the relationship to the cloud for more in-depth analysis, thereby relieving both the processing burden for the UAV devices and the bandwidth on the network. In addition to enabling local processing to enhance responsiveness, fog computing can also be used to further provide reliability, particularly in remote or isolated locations, without persistent broadband capabilities as a result of connectivity.

8.8.3 Distributed intelligence for remote areas

Distributed intelligence practice is particularly beneficial in an autonomous and remotely operated UAV to remotely control UAVs operations where secure and credible cloud processing is either unavailable or unsustainable. Decentralized or distributed multi-processor intelligence allows drones, edge devices, and local fog node intelligence to make on-board decisions at the point of collection, allowing for fast recovery, speed, and more distributed decision-making, all important aspects of autonomy when deployed in unpredictable environments with little or no communication. Environmental and remote UAV operations, such as search-and-rescue missions or environmental monitoring to assess human impacts in urban settings, can occur without real-time communications to other units or centralized control [27]. In addition, UAVs with distributed AI abilities can collaborate with other UAVs or local ground stations, including sensor devices where appropriate, to share information and improve mission outcomes. Collaborative units can operate effectively in constrained environments, even in less than ideal conditions related to environments with little or no communication to standard external local units. UAVs have no restraints and can participate effectively in a "swarm," yielding unpredictability and requiring coordinated dynamic multi-vector action based on shared information when necessary. Using swarm and distributed capabilities provides understanding at the systems level, enabling the UAV application to operate independently, cover more area, and mitigate potential misuse while avoiding unnecessary human remote interactions. Distributed and decentralized intelligence is crucial to reduce dependence on real-time communication and enable the UAV to operate continuously or with minimal degradation in environments where current cloud computing processes or systematic interactions from the operator or other agents can be limited or virtually incapacitated [28].

8.9 MULTI-UAV COORDINATION AND SWARM INTELLIGENCE

Multi-UAV Coordination and Swarm Intelligence is an area of study based on the multi-UAV concept of autonomous, cooperative operation, involving a cooperative effort among multiple uncrewed aerial vehicles (drones) to complete tasks that would be challenging or impossible to

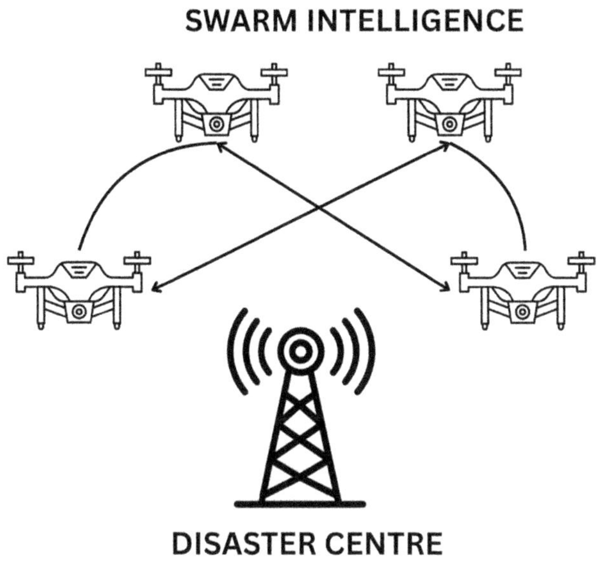

Figure 8.6 UAV assistance with swarm intelligence.

achieve using a single UAV [29]. Following nature's biological swarming behavior, such as the swarm behavior of fish, birds, or insects, UAVs can increase mission success, increase area coverage, and limit the risk of mission failure. This area draws upon swarm intelligence, and in this case, a decentralized, self-organizing system can emerge from the behaviors of single agents (drones) that simply follow a set of rules and swarm maneuvers as shown in Figure 8.6.

Multi-UAV coordination strives for distributed optimal resource allocation, task allocation, and decentralized decision-making that does not rely upon upper-level control, which can complicate operations and lower the resilience of the system [29]. Coordination of UAVs within a swarm occurs within the restrictions of multiple algorithms and methodologies, allowing for effective communication, navigation, and collaboration. These systems have to operate in dynamic environments and make changes to the mission plan, with real-time changes often being necessary to paths and actions. A common advantage of multi-UAV coordination is its systematic ability to cover large surfaces quickly and with redundancy, which is particularly advantageous to wide area surveillance and monitoring, search and rescue, and agricultural crop mapping. The emphasis on swarm intelligence arose due to the demand for more fully autonomous, scalable, and fault-tolerant UAV systems.

8.9.1 Swarm path planning algorithms

Swarm path planning algorithms enable cooperation among multiple UAVs to achieve a shared goal while maintaining efficiency, risk reduction, and obstacle avoidance. These algorithms leverage natural patterns, such as the flocking of birds or foraging of ants, to permit a group of autonomous UAVs to collaborate without centralized control [30]. Some examples include an artificial potential field to guide UAVs around obstacles while maintaining a flocking structure and decentralized algorithms, like the Consensus-based Bundle Algorithm (CBA) or the Optimal Reciprocal Collision Avoidance (CA) method. These techniques can handle evolving obstacles. Path planning for swarms is a crucial facilitator for applications such as search-and-rescue operations, environmental observation, and surveillance, necessitating a team of UAVs to collectively and effectively cover vast regions. Swarm path planning challenges involve ensuring communication, collision, performance, and reducing energy use.

8.9.2 Cooperative mapping and shared sensing

Cooperative mapping and shared sensing describe the teamwork of several UAVs to develop or revise a map of an area, especially when one UAV cannot effectively survey the region due to its size or intricacy [30]. Through the exchange of sensor information and processing power, the UAVs can deliver a more precise and current map of the area in real-time. Algorithms such as Simultaneous Localization and Mapping (SLAM) are commonly used in such efforts, where each UAV generates a map of a local area while simultaneously determining its own position. Each UAV maps the local area with its own sensors, and after mapping the environment, the information is shared with the other UAVs to combine to create a single global map of the environment all in real-time. Collaborative mapping and mapping in general can also be particularly useful in environments such as disaster relief, where UAVs need to collect the information quickly and possibly in low-visibility situations to map out the large area surrounding the disaster. A significant challenge is a seamless flow of data between UAVs to share the information and efficiently deploy the shared mapping and perception methods while minimizing the associated errors and time delays in UAV communication. Moreover, an efficient mapping system also requires optimization of the sensor platform, that is, cameras, LiDAR, etc., to not only map the environment but also to sufficiently perceive the environment in which the UAVs are mapping.

8.9.3 Collision avoidance systems

Collision avoidance systems are critical components to develop, since they allow UAV swarms to be both safe and autonomous, especially in denser environments or when there are other UAVs operating nearby. In most systems, the UAVs will rely on a mix of sensors, like radar, LiDAR, or cameras, which help the UAVs learn if an obstacle or another UAV is going to potentially occupy the same space. Often these types of systems will utilize navigation algorithms, such as the Reactive Navigation Algorithm (RNA) or Artificial Potential Field (APF) algorithms, that enable a UAV to navigate within a safe distance of obstacles while not colliding into those obstacles. Systems must also take into consideration dynamic obstacles that could be moving in twenty-first century airspace, along with adapting to a new state of dynamic obstacles. More advanced systems use predictive ML continuous models that allow for future collision risk evaluation and flight path adjustments in real time. The issue is designing systems for appropriate levels of response to avoid collisions while computationally having the transport type that can handle the fighting dynamics of a swarm volley [31]. In addition, we must also ensure that even with multiple UAVs present in constrained spaces, they can coordinate with each other, with a level of accuracy that does not interfere.

8.9.4 Distributed AI models

Distributed AI models designed for UAV swarms enable individual UAVs to make decisions and carry out actions independently but collaboratively. Models utilize decentralized learning (or decentralized reinforcement learning), whereby UAVs share perceptions of their surroundings and their actions to refine their decision-making [1, 5, 16]. Reinforcement learning (RL) and federated learning (FL) are often used, which enable each UAV to learn from its own experience as well as the experience of the entire swarm [32]. Distributed AI improves swarm performance on tasks such as task allocation, planning of a path, and resource management. Distributed AI can alleviate the need for centralized control, which enables the swarm to scale as well as become fault tolerant. The distributed systems do present some challenges, such as enacting consistency in the models, communications constraints, misinformation, or conflicting data. Distributed AI is particularly well suited for complex tasks such as environmental monitoring and military reconnaissance or precision agriculture, where decisions need to be made quickly by the swarm autonomously [32].

8.10 FUTURE DIRECTIONS AND RESEARCH OPPORTUNITIES

The future of research opportunities in the reports of drones, or UAVs, promises great potential for changing industries and increasing efficiency to grapple with complex global issues. As UAVs continue to become more complicated and integrated into our daily operations, they are a common technology tool, multi-modal, in fields such as agriculture, transportation, security, disaster relief, and more exciting research interests to address complexity, autonomy, safety/security, environmental impact, and the multi-faceted nature of systems. Listed below are several prominent sectors of research that will lead the future for UAV technologies.

8.10.1 AI-based autonomous disaster response

Autonomous disaster response systems based on artificial intelligence (AI) emphasize the use of UAVs to assess and respond automatically to natural disasters, including earthquakes, floods, and wildfires. UAVs can be deployed in such situations to collect critical data, locate survivors, deliver medical supplies, and contribute to infrastructure assessments and repairs, among other responsibilities. AI algorithms such as ML, computer vision, and deep learning can be applied to enhance the drone's capacity to navigate complex and unpredictable environments and make real-time decisions without human input. Another focus of research is swarm intelligence and UAV collaborative networks, where multiple UAVs can collaboratively synchronize their movements to cover a larger area with greater efficiency while decreasing response times. Future research will need to examine how to integrate AI with real-time data, such that autonomous UAVs can respond to changing conditions, for example, due to environmental hazards or communication interruptions. AI-based systems could also prioritize rescue missions or array targeting in some situations and optimize resource allocation for rescuers to save lives and minimize destruction.

8.10.2 Drone corridors for humanitarian

Drone corridors are established airspace areas intended to accommodate UAV deployments for use in disaster or emergency response areas safely. Drone corridors are established to keep UAVs away from a collision with other aircraft and ground risk while allowing for effective use of UAVs in humanitarian emergency events. The corridors would allow UAVs to transport supplies, evacuate victims, and survey or assess damage in areas where traditional transportation infrastructure was destroyed or could not be utilized. The research on the development of the drone corridor focuses on developing a real air traffic management system, monitoring UAVs during operations, and developing means by which UAVs can communicate and function safely in their designated airspace. Additionally, it will be important to establish regulations and policies to manage drone corridors in the national airspace, especially during an abundance of emergency operations. Establishing AV-safe routes that are integrated into the existing air traffic system could change the face of disaster organization and emergency response operations.

8.10.3 Emergencies

AI-based autonomous disaster response represents a developing research area focusing on the application of UAVs and AI technologies to enhance the efficacy and speed of disaster response operations. Following a natural disaster, such as an earthquake, flood, or wildfire, UAVs with advanced AI technology can quickly and autonomously navigate environments that are dynamic or hazardous, assess damage, and either deliver supplies or rescue individuals. Algorithms in AI, including computer vision, deep learning, and reinforcement learning, can detect survivors, assess infrastructure damage, and delineate a disaster-affected region in real-time. Swarm intelligence enables the multiple UAVs to work together to cover the whole area and minimize the response time. The difficulty lies in making sure that a system can function efficiently in severe conditions like communication, GPS signals, or sensor data might not perform as anticipated. Moreover, AI-powered systems will assist in distributing products and focus on rescue efforts, thereby preserving lives and minimizing effects during disaster response.

8.10.4 Quantum-secure UAV networks

As UAVs advance in critical infrastructure and sensitive areas, the security of communications and data transmissions becomes a growing concern. Quantum-secure UAV networks are an emerging area of cybersecurity that has the principle of quantum mechanics to protect communication channels from hacking. One opportunity for secure, tamper-proof data exchange between UAVs and ground stations is quantum key distribution (QKD), which is being investigated for UAV applications. Quantum encryption is based on the laws of physics, which provides a secure method for encryption that is secure from eavesdropping. It intercepts quantum encryption during transmission and detects intrusion. The urgency for quantum-secure communication will only rise as UAV networks expand in scope and complexity (military, government, and commercial applications) occur. Research efforts are focused on emerging quantum communication hardware and algorithms that have potential commercial utilities in UAV networks, including challenges in computational complexity, network scalability, and environmental obstruction.

8.10.5 Net-zero emission UAV designs

UAV designs focused on net-zero emissions strive to reduce, if not eliminate, the environmental impacts of UAV operations with a strong focus on sustainable energy sources and low-carbon technologies. The demonstrated carbon footprint of UAV fleets may be significant as they become more common, particularly in the fields of agriculture, logistics, and surveillance. Increasing UAV energy efficiency, creating lightweight materials, and implementing renewable energy harvesting systems (like solar panels) to increase flight time without releasing carbon dioxide are the main goals of net-zero design research. Furthermore, net-zero design will make UAV systems more sustainable through the design and substitution of unexplained materials in their manufacturing processes, reducing waste and energy expended in production. The overarching aim is for UAV systems to be environmentally sustainable, cost-effective, and usable without contributing to climate change as we continue the development of greener, low-carbon technologies.

8.10.6 Blockchain-based data integrity

Utilizing blockchain to ensure data integrity is becoming popular for guaranteeing veracity, security, and transparency of data collected by UAVs, especially in critical applications including military reconnaissance, environmental monitoring, and supply chain logistics. Particularly in industries where data integrity is critical, like healthcare or finance, it is imperative to ensure that this data is unchangeable, verifiable, and traceable. Without depending on any outside authority, blockchain technology offers a decentralized, resistant ledger for dependable data storage and verification. In the blockchain UAV network, the data generated by each UAV will be collected, recorded in blocks, and chained together in a distributed network of UAVs, offering an enormous barrier to altering the data undetected. The goal of this research is to optimize the use of blockchain with UAVs, including addressing storage concerns, computational costs, and transmission speed. Overall, blockchain exemplifies a trusted and transparent means of tracking and auditing UAV data, providing integrity to unmanned systems and more broadly total autonomous systems for high-risk applications.

8.11 CONCLUSION

UAVs are transforming a variety of fields with exciting and worthwhile applications to include disaster management and recovery, environmental monitoring, and data analysis. UAVs can provide whole-area situational awareness, can deliver supplies in disaster settings, and can assess damage in remote places, all of which has begun to incorporate UAVs into disaster-based organizational policies and practices. Using UAVs to explore and assess environments is shifting the research paradigm from anecdotal decision-making to a sustainable data-driven management tool to conserve and manage natural resources, e.g., surveying the forest to gauge pollution. Recent advancements and changes in UAV technology, including enhancements to sensors, assisted decision-making with AI and ML, are enabling a quicker and easier informed decision-making process for disaster or environmental applications. Edge, fog, and cloud computing offer extra real-time processing analytics designed to be scalable and enable data to be used effectively for both immediate and long-term impacts. The multi-UAV systems and swarm intelligence have significantly enhanced operational efficiencies in operations and facilitated the ability to complete complex tasks; however, they have also introduced additional considerations around topics such as cybersecurity, privacy, and ethics. As this type of technology continues to evolve, securing UAV networks and applying UAVs ethically within a humanitarian context will be key factors. Improving regulations and policy continues to develop internationally in recognizing the rapid advancement of UAVs, with several international standards already being developed to realize safety, privacy, and interoperability. As UAV operational capabilities grow, cross-national collaboration and data sharing will be essential in order to maximize the UAV usage potential. The future of UAVs in disaster management and environmental monitoring is looking very positive. The product such as AI-enabled autonomous disaster relief, quantum-secure UAV networks, and environmental monitoring is a very optimistic one. Innovations like AI-enabled autonomous disaster relief, quantum-secure UAV networks, and net-zero emission designs represent the latest possibilities of UAV adoption.

REFERENCES

[1] Mozaffari, M., Saad, W., Bennis, M., Nam, Y.-H., & Debbah, M. (2019). A tutorial on UAV's for wireless networks: Applications, challenges, and open problems. *IEEE Communications Surveys & Tutorials*, *21*(3), 2334–2360. https://doi.org/10.1109/COMST.2019.2902862

[2] Hong, L., Guo, H., Liu, J., & Zhang, Y. (2020). Toward swarm coordination: Topology-aware inter-UAV routing optimization. *IEEE Transactions on Vehicular Technology*, *69*(9), 10177–10187. https://doi.org/10.1109/TVT.2020.3003356

[3] Liu, Z., et al. (2023). Lightweight trustworthy message exchange in unmanned aerial vehicle networks. *IEEE Transactions on Intelligent Transportation Systems*, *24(2)*, 2144–2157. https://doi.org/10.1109/TITS.2021.3136304

[4] Ting-Ting, Z., Yan, C., & Ren-zhi, D. (2025). Autonomous decision-making of UAV cluster with communication constraints based on reinforcement learning. *Journal of Cloud Computing*, *14*, 12. https://doi.org/10.1186/s13677-025-00738-9

[5] Kong, X., Zhang, R., Chen, L., et al. (2023). Energy-aware path planning for long-endurance UAV's using constrained deep reinforcement learning. *IEEE Transactions on Vehicular Technology*, *72*(5), 5891–5904. https://doi.org/10.54097/gxy2w608

[6] Ryan, A., Zennaro, M., Howell, A., Sengupta, R., & Hedrick, J. K. (2004). An overview of emerging results in cooperative UAV control. In *43rd IEEE Conference on Decision and Control (CDC)* (Vol. 1, pp. 602–607). IEEE. https://doi.org/10.1109/CDC.2004.1428700

[7] Ekechi, C. C., Elfouly, T., Alouani, A., & Khattab, T. (2025). A survey on UAV control with multi-agent reinforcement learning. *Drones*, *9*(7), 484. https://doi.org/10.3390/drones9070484

[8] Giri, N., Swaroop, A., Virdee, B., Correia, S. D., & Polkowski, Z. (2026). Multi-agent reinforcement learning for complex decision-making. In A. Swaroop, B. Virdee, S. D. Correia, & Z. Polkowski (Eds.), *Proceedings of Data Analytics and Management (ICDAM 2025)* (Lecture Notes in Networks and Systems, Vol. 1600). Springer. https://doi.org/10.1007/978-3-032-03072-6_28

[9] Puente-Castro, A., Rivero, D., Pazos, A., et al. (2022). UAV swarm path planning with reinforcement learning for field prospecting. *Applied Intelligence*, *52*, 14101–14118. https://doi.org/10.1007/s10489-022-03254-4
[10] Qian, X., Shen, G., Lei, L., & Chen, Y. (2025). Multi-UAV cooperative search for moving targets: A deep reinforcement learning method. In Z. Yang & G. Sun (Eds.), *Proceedings of the 2nd International Conference on Networks, Communications and Intelligent Computing (NCIC 2024)* (Vol. 1360, pp. 1–9). Springer. https://doi.org/10.1007/978-981-96-5006-4_42
[11] Zhang, P. & Li, G. (2025). A cooperative dynamic target search approach for multi-UAV systems utilizing the MAPPO algorithm. *Discovery AI*, *5*, 153. https://doi.org/10.1007/s44163-025-00411-9
[12] Xu, Z., Wang, Z., Liu, R., et al. (2025). Efficient multi-UAV path planning in dynamic and complex environments using hybrid Multi-UAV optimization. *Journal of King Saud University–Computer and Information Sciences*, *37*, 125. https://doi.org/10.1007/s44443-025-00139-7
[13] Vishnoi, Vineet, Budhiraja, Ishan, Garg, Deepak, Garg, Sahil, Choi, Bong Jun, Shamim Hossain, M. (2025). Multiagent deep reinforcement learning based Energy efficient Information Classification: General resource management scheme for RIS assisted D2D users in 6G-aided smart cities environment, *Alexandria Engineering Journal*, 116, 609–620. https://doi.org/10.1016/j.aej.2024.12.004
[14] Anicho, O., Charlesworth, P. B., Baicher, G. S., et al. (2021). Reinforcement learning versus swarm intelligence for autonomous multi-HAPS coordination. *SN Applied Sciences*, *3*, 663. https://doi.org/10.1007/s42452-021-04658-6
[15] Jain, K. & Saini, H. K. (2023). An emergency rescue framework through smart IoT LPWAN. In *2023 International Conference on Advancement in Computation & Computer Technologies (InCACCT)*, Gharuan, India (pp. 668–672). https://doi.org/10.1109/InCACCT57535.2023.10141728
[16] Kankale, R., Paraskar, S., & Jadhao, S. (2021). Classification of power quality disturbances in emerging power system with distributed generation using space phasor model and normalized cross correlation. In *2021 8th International Conference on Smart Computing and Communications (ICSCC)* (pp. 340–345). IEEE. https://doi.org/10.1109/ICSCC51209.2021.9528195
[17] Zhou, T., Wang, L., & Wang, S. (2022). Distributed multi-UAV cooperative search and tracking based on deep reinforcement learning. *Aerospace Science and Technology*, *131*, 107989. https://doi.org/10.54097/gxy2w608
[18] Aldao, E., González-de Santos, L. M., Michinel, H., & González-Jorge, H. (2022). UAV obstacle avoidance algorithm to navigate in dynamic building environments. *Drones*, *6*(1), 16. https://doi.org/10.3390/drones6010016
[19] Jain, K. & Saini, H. K. (2023). An emergency rescue framework through smart IoT LPWAN. In *2023 International Conference on Advancement in Computation & Computer Technologies (InCACCT)*, Gharuan, India (pp. 668–672). https://doi.org/10.1109/InCACCT57535.2023.10141728
[20] Saini, H. (2025). GeNp-ODHR: Green energic and smart network performance estimation-based optimized deep hello routing for flying Ad Hoc network. *SAE International Journal of Connected and Automated Vehicles*, *8*(3), 319–335. https://doi.org/10.4271/12-08-03-0022

[21] Niedzwiecki, M. (1992). Multiple-model approach to finite memory adaptive filtering. *IEEE Transactions on Signal Processing*, *40*(2), 470–473. https://doi.org/10.1109/78.124965
[22] Yuan, H., Hao, H., Zhang, Y., Zhao, Y., & Chen, Y. (2022). Method for intelligently identifying underground safety accidents based on fully connected neural networks. In *2022 IEEE 6th Information Technology and Mechatronics Engineering Conference (ITOEC)* (pp. 367–370). IEEE. https://doi.org/10.1109/ITOEC53115.2022.9734567
[23] Zhou, T., Wang, L., & Wang, S. (2022). Distributed multi-UAV cooperative search and tracking based on deep reinforcement learning. *Aerospace Science and Technology*, *131*, 107989. https://doi.org/10.54097/gxy2w608
[24] Guo, J. et al. 2023. ICRA: An Intelligent Clustering Routing Approach for UAV Ad Hoc Networks. *IEEE Transactions on Intelligent Transportation Systems*, 24(2), 2447–2460. doi: 10.1109/TITS.2022.3145857
[25] Huang, F., Su, X., Wang, Q., et al. (2025). Physical property-aware multi-UAV spatial coordination for energy-efficient mobile edge computing. *Complex & Intelligent Systems*, *11*, 409.
[26] Zhou, J., Liu, M. L., Zhao, Y. F., et al. (2021). Energy-efficient UAV path planning based on double deep Q-network for mobile edge computing. *Computer Applications Research*, *38*(6), 1699–1703. https://doi.org/10.54097/gxy2w608
[27] Saini, H. K., & Jain, K. (2023, March 19). A novel deep energy efficient Hello packet scheduling for ad hoc networks in unmanned aerial vehicles. https://www.espublisher.com/journals/articledetails/843/
[28] Wan, K., Li, X., & Wang, Z. (2025). Cooperative localization and formation control of UAV's based on deep reinforcement learning under GPS-denied environments. *Acta Aeronautica et Astronautica Sinica*, *46*(2), 1–10. https://doi.org/10.54097/gxy2w608
[29] Guo, Y. & Liu, Z. (2024). UAV path planning based on deep reinforcement learning. *International Journal of Advanced Network, Monitoring and Controls*, *8*(3), 81–88. https://doi.org/10.2478/ijanmc-2023-0068
[30] Lv, Z., Xiao, L., Du, Y., Niu, G., Xing, C., & Xu, W. (2023). Multi-agent reinforcement learning-based UAV swarm communications against jamming. *IEEE Transactions on Wireless Communications*, 22(12), 9063–9075. https://doi.org/10.1109/TWC.2023.3268082
[31] Zheng, Z., Wei, C., & Duan, H. (2024). UAV swarm air combat maneuver decision-making method based on multi-agent reinforcement learning and transferring. *Science China Information Sciences*, 67, 180204. https://doi.org/10.1007/s11432-023-4088-2
[32] Cheng, C. (2025). Application of Reinforcement Learning to UAV Decision Making and Path Planning. *Frontiers in Computing and Intelligent Systems*, *14*(1), 30–33. https://doi.org/10.54097/gxy2w608
[33] Kaliappan, V. K., Nguyen, T. A., Jeon, S. W., Lee, J. W., & Min, D. (2023). Deep multi-agent reinforcement learning-based decentralized swarm UAV control framework for persistent surveillance. In S. Lee, C. Han, J. Y. Choi, S. Kim, & J. H. Kim (Eds.), *The Proceedings of the 2021 Asia-Pacific International Symposium on Aerospace Technology (APISAT 2021)* (Lecture Notes in Electrical Engineering, Vol. 913). Springer. https://doi.org/10.1007/978-981-19-2635-8_70

Chapter 9

Low-carbon and energy-aware UAV operations for sustainable smart mobility ecosystems

Bebesh Tripathy, Riya, and Hemant Kumar Saini

9.1 INTRODUCTION

Unmanned aerial vehicles (UAVs) are all the rave of the businesses of tomorrow due to the pace of their implementation in logistics, environmental surveillance, accurate agriculture, and intelligent mobility. However, the increasing demand for the operations of the UAV fleets exposes new concerns about their energy consumption and environmental sustainability. As the world's industries become low-carbon and energy-efficient, more than ever, energy-efficient UAV ecosystems are in demand. Further sustainability conceptualizations, including renewable energy integration, green technologies, and life cycle carbon reduction strategies (Huang et al., 2025), reveal the fundamental concepts for the reconsideration of the UAV operative models in relation to the climate-neutral objectives (Habib et al., 2025; Shaheen & Khan, 2026). The energy consumption of the UAV device is a complicated problem and is associated with how the propulsion system is designed, the chemistry of the battery, the time of the proposed flight, the weather, and the efficiency of the route. Overall, such problems are similar to the large problems existing in other industries, e.g., building systems, industrial processes, and transport electrification. As an example, a study of operational emissions in residential and industrial energy systems has suggested that optimization of energy systems could significantly reduce emissions during operation and infrastructure life cycle (Cheng et al., 2025; Song et al., 2026). Similarly, drone operations require a more comprehensive calculation of emissions, including such aspects as battery production, recharging, and disposal. The shift to a low-carbon economy, with the foundations of the research regarding life cycle carbon footprint in fashion, construction, and mobility systems, highlights the fact that the integration of sustainability should be considered throughout the design to deployment (Wu et al., 2025; Jeleniewicz et al., 2025). Individual innovations in clean and energy-efficient advanced energy systems and materials that are environmentally friendly may make a feasible change in using UAVs that have a sustainable clean-energy source. Scientific efforts in a hydrogen-powered aerial vehicle, renewable-powered microgrids, and hybrid energy

 DOI: 10.1201/9781003731580-9

technologies are indicating the tremendous reduction of emissions during operation (Li et al., 2026; Liu, 2025). Meanwhile, on industrial scales to a significant degree, they are also deployed to transportation systems and smart energy scheduling; similarly, optimization algorithms can also provide a sound basis for efficient UAV routing and the mission-planning frameworks (Zhang et al., 2026; Mencaroni et al., 2025). These solutions are an indication that energy demanded by UAVs can be reduced dramatically as long as one applies optimization models that are based on information, take into account various goals, and reduce the amount of energy used, maximizing operational performance as much as it can be. In addition to that, the continued deployment of digital solutions (that comprise blockchain, AI, and approximate computing) can provide new viable situations for the real-time capture of information and carbon accounting in UAV operations. Excellent examples of traceability and decentralized decision-making as the primary contributors to the reduction of emissions on scale are the use of blockchain-based logistics assessments of electric vehicles and carbon-conscious manufacturing (Mofatteh et al., 2026; Vitali et al., 2026). Alternatively, AI-based predictive models applicable to infrastructure, energy systems, and industrial planning missions can serve as good role models to be followed in the UAVs industry through intelligent routing, predictive maintenance, and predictive energy consumption (Wójcik et al., 2026; King, 2025). The initiative toward low energy consumption and low-carbon returns on UAV systems correlates with global sustainability commitments, the idea of a circular economy, and available technological solutions that assist in the sustainable development goals (SDGs). The current research on infrastructures not associated with carbon emissions, green technologies, progressive integration of renewable energy, and optimization of life cycle is an informative resource in the direction of the next generation of clean UAV systems that produce fewer carbon gases but are effective in terms of their activities (Giama et al., 2025; Saini, 2024). The main content of sustainable deployment of UAVs should be energy efficiency, eco-design, and reducing the carbon footprint, which will become realizable in all market spheres and all types of environments because the use of UAVs will continue to increase.

9.1.1 Role of UAVs in sustainable mobility and SDG goals

UAVs are becoming more widely recognized as key components that enable sustainable mobility. This is, first of all, due to their ability to facilitate low-emission logistics, give a quick response to the service delivery, and open the way for efficient urban mobility solutions. In fact, UAVs are very much in line with the sustainable development goals (SDGs) such as SDG 9 (Industry, Innovation, and Infrastructure), SDG 11 (Sustainable Cities and Communities), and SDG 13 (Climate Action), as they allow for less reliance on the conventional transportation systems that emit a lot of carbon. Green finance, decentralized energy systems, and low-carbon transitions research provide evidence that the adoption of new technological pathways has an overall positive effect on the climate-neutral mobility frameworks. Among the vehicles, UAVs are the ones that can directly contribute by optimizing aerial operations (Habib et al., 2025). What is more, the chosen sustainable materials and carbon-aware scheduling in industrial processes tell us the importance of the eco-efficient design principles that have to be deeply embedded in the UAV ecosystems if one wants to talk about sustainability in the future (Varghese et al., 2026; Mencaroni et al., 2025). Besides that, recent research on renewable energy infrastructure, hydrogen-powered engines, and distributed microgrids indicates that the UAV fleets can be integrated into green energy infrastructures, thereby rendering the cutback of operational emissions more practical. The use of UAVs in clean-energy infrastructure goes a long way in the establishment of low-carbon industrial facilities and energy-efficient built environments. Both of them are systemic solutions to sustainable mobility (Giama et al., 2025; Cheng et al., 2025). Due to the transformations that are occurring in the global industries, they are shifting to climate and digitally enabled models of transportation. UAVs can also introduce a bit of fresh air in this transformation, as there is a chance of having the capability to operate with lower energy requirements, disaster resiliency, and facilitation of easy monitoring of the environment, which is facilitated by data collection. These are the reasons that make UAVs not only the means of mobility but also the strategic contributors to the achievement of the global sustainability goals and the charting of the paths that lead to climate mitigation (Jain, 2025; Zheng et al., 2025).

9.1.2 Environmental challenges in UAV operations

Even when operated with sustainability in mind, UAVs remain a cause for environmental concerns. This is mainly because of high energy consumption, less than optimal battery efficiency, and carbon emissions that are associated with supply chain processes. For example, the manufacturing, disposal, and frequent charging of batteries generate carbon emissions indirectly, which is becoming a challenge in the electric mobility and renewable energy sectors as well. Comparing the carbon footprint of UAVs with that of lithium-ion energy storage and industrial operations, if the former is not integrated into a low-carbon infrastructure, the latter can become the main source for the carbon footprint of UAVs (Sadavarte et al., 2025; Hofstetter et al., 2025). On top of all these, the resource-heavy UAV manufacturing process also raises the issues of metal extraction and non-metals, not to mention electronic components, all pointing to an increase in life cycle emissions in the future (Farazmandnia & Ilinca, 2025; Jeleniewicz et al., 2025). The effect of UV noise, the impact on nature, and the souls of the weather-net and their effect on the environment are seldom the focus of the traditional airborne monitoring systems. Research on carbon-intensive infrastructures such as tunnels, buildings, and energy systems—which are heavily influenced by environmental conditions—shows that in such cases, operations need to devise energy and carbon management strategies for adaptation (Song et al., 2026; King, 2025). To operate in such a volatile atmosphere is to have an increased UV power consumption level and thus a shorter flight duration potential. Again, poor planning of the route, overuse of the drone in a certain position, and the inefficiency of the mission in general are the reasons behind energy wasting and, hence, the need for optimization solutions. On the one hand, as the number of drones to be hit increases, their environmental constraints must be addressed systematically to avoid their unintended ecological and carbon impacts (Bertone & Manzone, 2025; Al-Osaimi et al., 2025).

9.1.3 Need for energy efficiency and carbon reduction

Energy-efficient UAV operations are a must if the world wants to meet the rising global decarbonization targets that have been put forward. Research in industrial sectors has revealed time and again that energy optimization combined with low-carbon planning is a winning strategy, as it drastically reduces both greenhouse gas emissions and operating costs (Liu, 2025; Ma et al., 2025). Similarly, such strategies may be used to achieve significant improvements in both endurance of flight and performance in carbon in the case of more advanced energy scheduling, multi-objective optimization, and the most efficient propulsion designs. The insights about the low-carbon logistics, hydrogen manufacturing, and renewable microgrids suggest that UAVs can also become very green when incorporated into a clean-energy ecosystem (Bae & Park, 2025; Marsigny et al., 2025). Adding to that, the usage of AI-based algorithms, life cycle analyses, and energy-conscious online platforms can also establish the path toward energy-saving practices and, at the same time, make sure that the resources can be utilized in a responsible manner. The studies of green computing, decentralized energy management, and approximate computing also indicate the benefits of smart systems that are capable of operating the distribution of power in an optimal manner and, at the same time, ensuring that the emission is minimal (Rafique et al., 2026; Vitali et al., 2026). Using the knowledge in the area of UAV operations, companies will be capable of developing adaptable routing solutions, predictive-based maintenance, and cognizant energy requirement fleet coordination. The problem of energy and carbon performance of UAVs must be central to the discussion in case we do take a leap toward the development of the robust, low-emission, net-zero-focused aerial ecosystems that industries must have so that they can cluster their net-zero goals and environmentally aware mobility solutions (Saini & Jain, 2023, March 19).

9.2 LITERATURE REVIEW

The increasing amount of literature on advanced wastewater treatment technologies indicates the innovation toward sustainability and resource recovery. Chen et al. (2026) report on algal-bacterial granular sludge treatment as a combined approach to polish and recover nutrients in wastewater with particular attention to technical aspects of scale-up. The paper by Hong et al. (2025) offers a materials-focused breakthrough in the form of Ir-doped $NiFe_2$ O_4/TiO_2 heterojunction anodes that are capable of decentralized saline wastewater treatment and simultaneous hydrogen production and show synergies between treatment and renewable energy generation. These articles demonstrate the tendency to balance treatment with energy or resource valorization to enhance the overall life cycle performance (Chen et al., 2026; Hong et al., 2025). Table 9.1 shows a summary of all the literature in tabular format.

A number of works analyze the technological directions of decarbonizing the industry and power systems, such as hydrogen integration and the use of carbon capture. Li et al. (2026) examine the issue of hydrogen injection in blast furnaces, including the impact of injection temperature, although Marsigny et al. (2025) also put forward operating solutions of direct reduction of iron ore by hydrogen in shaft furnaces—both papers suggest operating levers to reduce CO_2 in the steel sector. Siskos et al. (2026) compare the production of methanol using cement CO_2 emissions and provide techno-economic insights into the use of industrial CO_2 streams. All these works trace feasible pathways to industry decarbonization via hydrogen and carbon use plans (Li et al., 2026; Marsigny et al., 2025; Siskos et al., 2026). Another significant cluster is transportation, mobility, electrification, and lifetime influences. Sadavarte et al. (2025) contrast the U.S. climatic and air pollution footprints of the Li-ion BEVs and ICE vehicles in the U.S. by comparing them, expressly accounting for the systemic energy system responses, and Mofatteh et al. (2026) examine the viability of EV adoption in outbound logistics using a blockchain-based life cycle sustainability assessment. Kang et al. consider the topic of EV charging emissions under the conditions of demand uncertainty and grid constraints, which is significant because it depends on the integration of both charging strategy and grid planning. Collectively these publications highlight that the advantages of vehicle electrification are only heavily contingent on the context of the energy system, life cycle considerations, and logistics design (Sadavarte et al., 2025; Mofatteh et al., 2026). The research on energy systems modeling and optimization offers the frameworks to make the tradeoffs of cost and reliability against emissions in the multi-energy scenario. Zhang et al. (2026) introduce the large-scale stochastic production decision framework of coupled economy-environment-energy systems, and Ma et al. (2025) also introduce the multi-objective, distributionally robust low-carbon planning model of the integrated multi-energy microgrids that also utilize real-time workload

Table 9.1 Summary of key sustainability and low-carbon transition studies

Ref no.	*Title*	*Author and year*	*Findings*	*Research gaps*
1	Algal-bacterial granular sludge process for sustainable wastewater treatment	Chen et al. (2026)	Enhanced nutrient removal efficiency and improved granule stability through algal-bacterial integration; demonstrates potential for sustainable wastewater treatment.	Large-scale pilot demonstrations needed; long-term granule stability uncertain; light–microbe interaction optimization required.
2	Green finance and environmental decentralization drive OECD low-carbon transitions	Habib et al. (2025)	Green finance and decentralization significantly accelerate low-carbon transitions in OECD regions using empirical econometric evidence.	Limited applicability to developing economies; long-term financial return dynamics unexplored.
3	Ir-Doped $NiFe_2O_4/TiO_2$ heterojunction anode for saline wastewater treatment and H_2 production	Hong et al. (2025)	High-efficiency anode enables simultaneous saline wastewater treatment and hydrogen generation; offers scalable clean-energy co-benefits.	Need for real-world industrial testing; cost and scalability challenges persist.
4	Subsurface temperature dynamics in borehole heat exchanger systems	King (2025)	Provides long-term thermal behavior data to optimize geothermal system performance and improve fluid flow modeling.	Limited extreme climate testing; generalized applicability across geological conditions not validated.
5	Minimizing desflurane use via education and vaporizer removal	Lehmann et al. (2025)	Educational interventions and vaporizer removal significantly reduce desflurane usage, lowering healthcare carbon footprint.	Long-term behavioral adherence unclear; lacks economic evaluation; scalability across hospitals not assessed.

(*Continued*)

Table 9.1 (Continued)

Ref no.	*Title*	*Author and year*	*Findings*	*Research gaps*
6	Hydrogen injection in blast furnaces	Li et al. (2026)	Injection temperature strongly influences hydrogen reduction efficiency and operational stability; can meaningfully reduce CO_2 emissions in steelmaking.	Real-world operational variability not fully explored; hybrid decarbonization pathways not examined.
7	Blockchain-based life cycle assessment for EV logistics	Mofatteh et al. (2026)	Blockchain improves transparency and accuracy in EV life cycle sustainability assessments; EV adoption lowers logistics emissions.	High blockchain computational overhead; real-world pilot validations needed; battery supply chain uncertainty persists.
8	Green IT technologies for reducing carbon footprint	Rafique et al. (2026)	Identifies IT optimization methods such as virtualization, efficient cooling, and cloud optimization that reduce IT carbon emissions.	Lack of comparative quantification; hardware life cycle footprint not evaluated; limited real-time optimization techniques.
9	Footprint comparison of BEVs vs. ICEs	Sadavarte et al. (2025)	BEVs outperform ICEs in climate impact under low-carbon power mixes; systemic modeling captures upstream energy dynamics.	Recycling impacts of batteries not integrated; regional grid variability requires higher-resolution modeling.
10	Sustainable stratagems and financial performance	Shaheen & Khan (2026)	Sustainability initiatives positively influence financial performance, showing strategic and economic alignment.	Lacks industry-specific breakdown; short-term vs long-term effects unexplored.
11	Methanol production from cement CO_2 emissions	Siskos et al. (2026)	Demonstrates viability of methanol production from industrial CO_2, offering feasible circular-carbon utilization pathways.	Needs real flue-gas validation; catalytic degradation under real conditions unclear; energy cost sensitivity unassessed.

responses of data centers. Liu (2025) concentrates on power-to-X and carbon capture technologies that may have multi-layered schedules in the decarbonized urban areas. These methodological improvements are becoming more sophisticated in the way they address uncertainty and real-time interaction between energy demand (including data centers) and options with low-carbon supply (Zhang et al., 2026; Ma et al., 2025; Liu, 2025). Circularity and life cycle carbon reduction in the built environment and construction materials literature. Varghese et al. (2026) study the topic of geopolymer concrete using industrial byproducts to lower building emissions, and Jeleniewicz et al. (2025) compare timber and steel buildings in terms of environmental impact and circularity in residential construction. Cheng et al. (2025) include carbon footprint optimization as part of building life cycle cost analysis, and Song et al. (2026) make carbon footprints of road tunnels regionally—these studies all point to materials selection, design optimization, and spatial-temporal modeling as decarbonization levers of the built environment (Varghese et al., 2026; Jeleniewicz et al., 2025; Cheng et al., 2025). Research on renewable integration, storage, and powertrain design focuses on system-level emissions. Giama et al. (2025) optimize building heat pumps with photovoltaic thermal (PV/T) systems to achieve carbon neutrality, and Hofstetter et al. (2025) optimize the design of electric powertrains with multi-objective design optimization to reduce carbon footprints. The microgrid planning work done by Ma and his colleagues (Ma et al., 2025) connects distributed resources and data-center demand response. These contributions show that renewables and electrification can be extended to generate emission savings, but coordinated system scheduling and hardware design are needed (Giama et al., 2025; Hofstetter et al., 2025; Ma et al., 2025). Sustainability, circular economy, and materials-based research touch on the end-of-life and reuse. Farazmandnia and Ilinca (2025) provide a life cycle assessment of recyclable wind turbine blades, whereas Ouro Miguélez et al. (2025) evaluate demountable and reconfigurable steel structures. Kothari et al. (2025) explore the CO_2 sequestration through carbide slag mineral carbonation and valorization of carbonated product, which represents a circular opportunity at the industrial scale. These articles refer to design-to-recycling and material-loop closure as effective measures to lower the impact of the life cycle (Farazmandnia & Ilinca, 2025; Ouro Miguélez et al., 2025; Kothari et al., 2025). Sectoral and operational healthcare, laboratory, and hospitality pathways to emissions reductions are new trends. Lehmann et al. (2025) study theater and anesthetic practices in hospitals; the former one (the former) defines the directions that net-zero operating theaters should follow, whereas the latter one (the latter) shows that desflurane can be decreased through specific interventions (education and removing vaporizers). The recommendations discussed by Arnoldo et al. (2025) offer the best environmental activities that can be applied to clinical laboratories, which offer short-term operational solutions. These practical investigations demonstrate that high-impact and cheap operational modifications can be made in the medical facilities (Lehmann et al., 2025; Arnoldo et al., 2025). Emission-conscious computing, information technology, cloud computing, and artificial intelligence are discussed both in terms of control and application. Vitali et al. (2026) suggest adaptive green cloud applications, which trade in emissions, revenue, and user experience, with approximate computing. In a survey of Green IT technologies to minimize IT carbon footprints, Rafique et al. (2026) and in carbon-aware scheduling to reduce manufacturing emissions, Mencaroni et al. (2025) present their results. Such studies signify that software-based controls and scheduling policies are potent additions to the hardware decarbonization endeavors (Vitali et al., 2026; Rafique et al., 2026; Mencaroni et al., 2025). The methodological and policy analyses cut across finance, governance, and life cycle assessment strategies. According to Habib et al. (2025), green finance and environmental decentralization are associated with OECD low-carbon transitions, which implies the importance of institutional and financial incentives. Hebala et al. (2025) discuss energy management strategies used in the higher education sector to reduce the footprint, focusing on the organizational strategies. Liu (2025) and Zhang et al. (2026) are examples of sophisticated modeling methods of policymaking and system planning that consider tradeoffs between the economy and the environment (Habib et al., 2025; Hebala et al., 2025; Liu, 2025; Zhang et al., 2026). Application-oriented life cycle analyses (LCA) as well as various case studies unfold local, situation-dependent realities that show the different sectors' behavior. Wu et al. (2025) analyze the environmental impact of a luxury fashion brand's fast fashion event booth in a closed-loop system, while Cheng et al. (2025) perform carbon footprint reduction optimization for a MEGA house case study. Klarenbeek et al. (2025) adapt methods of healthcare FMEA to localize surgical waste in the orthopedic department, whereas Nayek and Das (2025) solve main pumping problems for lower energy and capital costs. These local investigations highlight the potential of locally adjusted LCA and optimization that can uncover real interventions with measurable impacts (Wu et al. 2025; Cheng et al., 2025; Klarenbeek et a., 2025; Nayek & Das, 2025). Several contributions along this line of research have also been taken to the next level by combining emerging materials, manufacturing, and AI applications for the conservation of natural resources. Wójcik et al. (2026) compile a review of AI/ML tools in regeneration and control directed to industry, while Wójcik et al.'s (2026) perspective paper serves as a linkage between works on process-level optimization in manufacturing and powertrain design (Wójcik et al., 2026; Hofstetter et al., 2025). Bae and Park (2025) study the H_2-DRI technique with solid and liquid carbon byproducts under electrified smelting scenarios, thus extending the research on carbon-emission abatement in metallurgy. The examined research works contribute to an interdisciplinary trend—the combination of AI, material innovation, and process redesign that can pave the way for net-zero manufacturing processes (Wójcik et al., 2026; Bae & Park, 2025).

9.3 CONCEPTUAL FRAMEWORK

The proposed model of energy-efficient operations of UAVs is based on the combination of a sustainable design, sophisticated optimization strategy, and a life cycle-friendly approach to carbon management. The energy consumption of UAVs depends on the hardware technologies, aerodynamic parameters, the mission constraints, and the conditions in the environment; hence, it is necessary to introduce a profound framework with references to the proficient energy systems, life cycle analysis, and environmental optimization schemes. The studies in industry and infrastructure have proven that the life cycle thinking in the system design can contribute greatly to emissions reduction and improved performance of the system (Cheng et al., 2025; Song et al., 2026). When these principles are transferred to UAV ecosystems, they can give order to the evaluation of energy consumption at the phases of manufacturing, deployment, and disposal to make sure that it corresponds to the objectives of the circular economy and sustainability (Wu et al., 2025; Jeleniewicz et al., 2025). Anthropomorphic to the conceptual model is the effort of AI and machine learning to streamline UAV flight operations, resource distribution, and foreseeable maintenance (Figure 9.1).

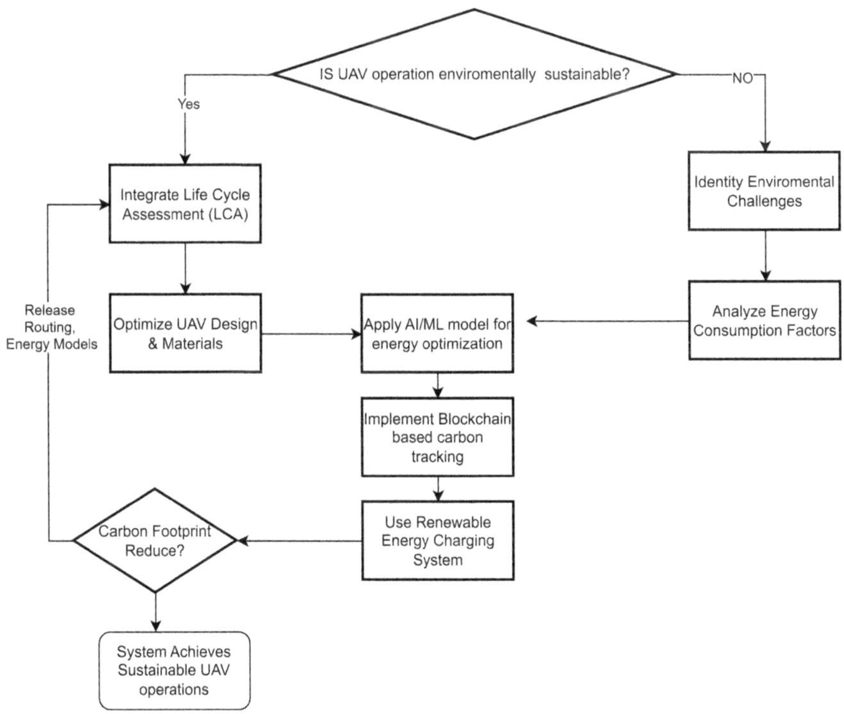

Figure 9.1 Conceptual framework for energy-efficient and low-carbon UAV operations.

Reinforcement, multi-objective, and predictive models are all machine learning methods that have effectively improved efficiency in industrial systems, logistic networks, and smart manufacturing (Wójcik et al., 2026; Zhang et al., 2026). These strategies, when used with UAV systems, make it possible to dynamically route, manage energy in real-time, and make decisions responsive to environmental factors. It is also true that AI models are able to mimic energy-intensive conditions, predict power losses, and determine the most efficient mission settings, which is similar to other energy-intensive areas of the economy, including electric mobility and integrated microgrids (Liu, 2025). Other emerging digital technologies that are included in the framework are blockchain and approximate computing, which increase transparency, accountability, and energy efficiency in UAV ecosystems. Life cycle tracking based on blockchain facilitates decentralized carbon accounting, energy traceability, and verifiable emissions reporting, which are already established in sustainable logistics and industrial carbon management (Mofatteh et al., 2026; Vitali et al., 2026). Similarly, green computing plans lower the computational load and optimize the efficiency of onboard UAV processing, as it helps in lowering the energy consumption in the course of mission execution. A combination of these technologies helps a system where UAV fleets serve in an accountable manner, with lesser processing overhead and enhanced environmental performance (Rafique et al., 2026; Giama et al., 2025). Lastly, the conceptual framework locates renewable energy integration and eco-efficient infrastructure as the key pillars toward sustainable UAV operations. Research on hydrogen-powered fuel, renewable-powered charging systems, and carbon-neutral energy scheduling focuses on the significance of integrating operational UAV systems with clean-energy sources (Bae & Park, 2025; Giama et al., 2025). Solar, wind, or hydrogen microgrid-based UAV charging stations will have the ability to significantly decrease life cycle emissions while providing fleets with permanent availability. In addition, industrial and transportation planning built on carbon-conscious scheduling strategies offer a guide on UAV mission scheduling that avoids peak energy demands and integrates the operations with the low-carbon energy windows (Mencaroni et al., 2025; Ma et al., 2025). A combination of these aspects offers a complete framework for designing the next-generation UAV ecosystems that are not only energy-efficient and responsible to the environment but also in full compliance with the global sustainability goals.

9.4 FUNDAMENTALS OF UAV ENERGY CONSUMPTION

The energy consumption of UAV is mostly dependent on the aerodynamics of the vehicle, the propulsion system, the weight of the cargo, the conditions of the environment, and the nature of the power system. The electric propulsion system is among the primary energy demand sources, and battery efficiency, discharge rates, and temperature sensitivity are the primary drivers of the overall endurance of the flight. These parameters are quite comparable to those that form the key obstacles in electric mobility and renewable energy systems, where the operation of systems is determined by the battery chemistry, variability of loads, and real-time energy demands (Sadavarte et al., 2025; Hofstetter et al., 2025). Moreover, the production of the UAV involves materials and components that are not only energy-intensive processes but also lead to the indirect increase of the emissions of the entire life cycle. This is similar to the energy dynamics of industrial production and built-environment systems, where the choice of materials and the structural design have a significant impact on energy performance (Farazmandnia & Ilinca, 2025; Jeleniewicz et al., 2025). Therefore, knowledge of the topic of UAV energy consumption demands an integrated approach that takes into account not only operational energy but also embedded energy in hardware and materials. The facts of nature—wind resistance, air density, altitude, and temperature—are decisive factors of the real-time UAV energy consumption. Table 9.2 shows the fundamentals of UAV energy consumption.

The flight habits, such as hovering, rapid acceleration, and changing altitude, also contribute to the power consumption, and these are comparable

Table 9.2 Fundamentals of UAV energy consumption

Factor	*Description*	*Impact on energy consumption*	*Typical values/notes*
Propulsion system	Electric motors, propellers, ESC efficiency	Determines baseline power demand	60–70% of total power draw during flight
Battery characteristics	Battery type, capacity, discharge rate, degradation	Influences endurance and charging frequency	Li-ion manufacturing emits 60–100 kg CO_2-eq per kWh; degradation after 150–250 cycles
Payload weight	Cameras, sensors, packages	Increased weight → higher thrust required → more energy	Every additional 100 g increases power consumption by 4–8%
Aerodynamics	Airframe shape, drag, rotor design	Poor aerodynamics cause higher energy loss	Aerodynamic improvements can reduce energy use by 12–20%
Flight mode	Hovering, climbing, cruising	Different modes require different power levels	Hovering increases energy use by 20–35%; climbs require peak power

(Continued)

Table 9.2 (Continued)

Factor	*Description*	*Impact on energy consumption*	*Typical values/notes*
Environmental conditions	Wind, temperature, humidity, air density	Affects thrust, stability, and battery performance	Headwinds increase demand by 10–15%; low temperatures reduce battery efficiency by 10–20%
Routing and mission profile	Path planning, speed, altitude, maneuvers	Directly impacts energy expenditure	Optimized AI routing reduces energy use by 18–27%
Component efficiency	Electronics, sensors, computing load	High computation draws internal power	Green computing reduces losses by 10–15%
Manufacturing and material footprint	Embedded energy in materials and production	Indirect life cycle energy impact	Material-related life cycle emissions account for 30–50% of total

with the energy variations in tunneling operations, building systems, and large-scale industrial processes, which require constant accommodation of environmental conditions (Song et al., 2026; King, 2025). On top of that, the impact of the mission on the energy usage of the UAV, such as the planning of the route, the allocation of the payload, and the duration of the operations, can be substantial. The optimization of industrial scheduling and microgrid management leads to energy-saving strategies; this is where the planning of energy use plays a key role, as it cuts down on wasted energy (Mencaroni et al., 2025; Liu, 2025). These facts lead to the collective acknowledgment that fundamental UAV energy consumption is a multidimensional issue, influenced by design, environment, mission dynamics, and technological infrastructure. Figure 9.2 shows the layered vertical flowchart of fundamental UAV energy consumption factors.

9.4.1 Carbon footprint of UAV operations

The carbon footprint of UAV operations has been influenced by the direct operational emissions and indirect life cycle impacts resulting from battery production, component manufacturing, energy-consuming charging, and end-of-life waste. Though electrically powered UAVs do not emit any tailpipe emissions, life cycle assessments reveal that the production of one kWh of lithium-ion battery capacity can release 60–100 kg CO_2-eq, depending on the source of the material and the method of production (Sadavarte et al., 2025) (Table 9.3).

Corroborating evidence from renewable energy and built-environment research reveals that the embedded carbon in materials is responsible for 30–50% of the total life cycle emissions (Jeleniewicz et al., 2025). Moreover,

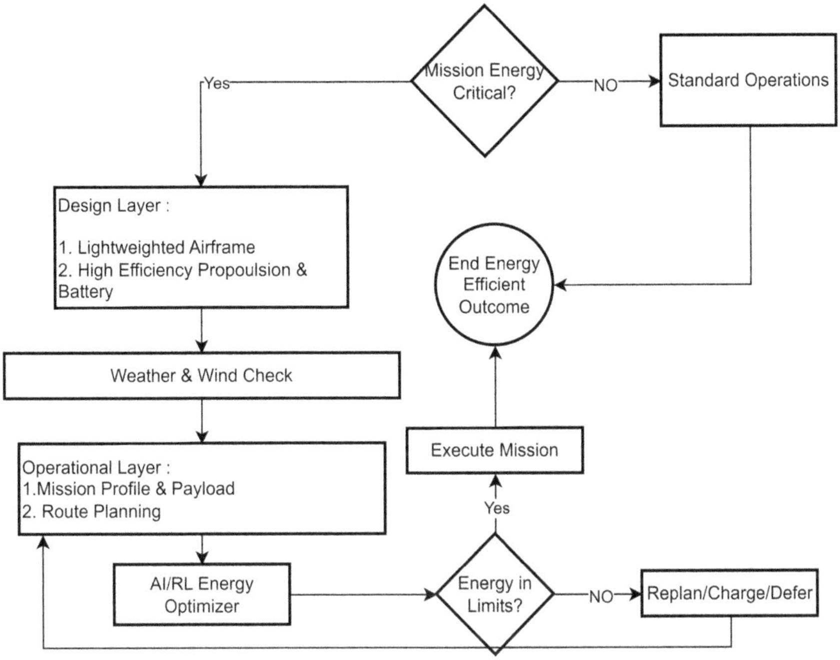

Figure 9.2 Layered vertical flowchart of fundamental UAV energy consumption factors.

the carbon emissions of UAVs during operation depend on the carbon intensity of the electricity used for charging; charging UAV fleets from grids powered by fossil fuels can increase CO_2 emissions from operations by 25–40%, thus following the same trend as energy-intensive industrial systems (Hofstetter et al., 2025; Cheng et al., 2025). Figure 9.3 shows the carbon footprint contribution of UAV operations.

Besides production and charging, the carbon impact of UAVs depends on their mission and the surrounding environmental conditions. For example, the power draw can be increased by 20–35% due to hover, while the energy consumption can be raised by more than 15% if the UAV flies against the strong winds because of aerodynamic drag—the environmental sensitivity here is similar to that which is reported in tunnel and building energy studies (Song et al., 2026; King, 2025). Deep discharge cycles on batteries accelerate battery degradation, thus shortening replacement intervals to 150–250 cycles and, thereby, contribute more emissions from battery disposal (Farazmandnia & Ilinca, 2025). Furthermore, when inefficiencies due to routing and payload handling issues are combined with fleet operations, the energy wasted can be 10–25% more than that of industrial scheduling and carbon-aware manufacturing systems, where a similar pattern of inefficiencies has been documented (Mencaroni et al., 2025; Liu, 2025).

Table 9.3 Carbon footprint of UAV operations

Carbon source/factor	*Description*	*Impact on emissions*	*Typical values/numerical data*
Battery manufacturing	Emissions from producing Li-ion batteries used in UAVs	Major contributor to life cycle footprint	**60–100 kg CO_2-eq per kWh** of battery capacity
Battery degradation and replacement	Frequent replacement due to cycle limits increases embedded carbon	Raises life cycle emissions	Degrades after **150–250 charge cycles**, raising replacement frequency
Electricity for charging	Depends on grid carbon intensity	Increases operational CO_2 if a fossil-powered grid used	Fossil-dominant grid increases emissions by **25–40%** per charge cycle
Material production	Manufacturing of carbon fiber, aluminum, motors, electronics	Adds indirect (embedded) emissions	Material-related CO_2 forms **30–50%** of life cycle emissions
Mission profile and flight behavior	Hovering, rapid climbs, aggressive maneuvers	Raises per-flight energy demand	Hovering increases consumption by **20–35%**; climbs cause spikes of **15–22%**
Environmental conditions	Wind, temperature, humidity	Alters aerodynamic drag and battery efficiency	Headwinds increase energy use by **10–15%**; cold reduces battery efficiency by **10–20%**
Payload mass	Higher weight requires more thrust	Raises total per-mission carbon emissions	Every **100 g** added → **4–8%** increase in energy demand
Inefficient routing	Longer or poorly optimized paths	Waste energy and increase emissions	Suboptimal routing increases emissions by **10–25%**
End-of-life disposal	Battery and component waste management	Causes additional carbon and ecological impact	Improper battery disposal adds **8–12%** extra life cycle emissions
Charging infrastructure	Non-renewable charging stations	Raises carbon footprint significantly	Renewable-powered charging reduces emissions by **32–45%**

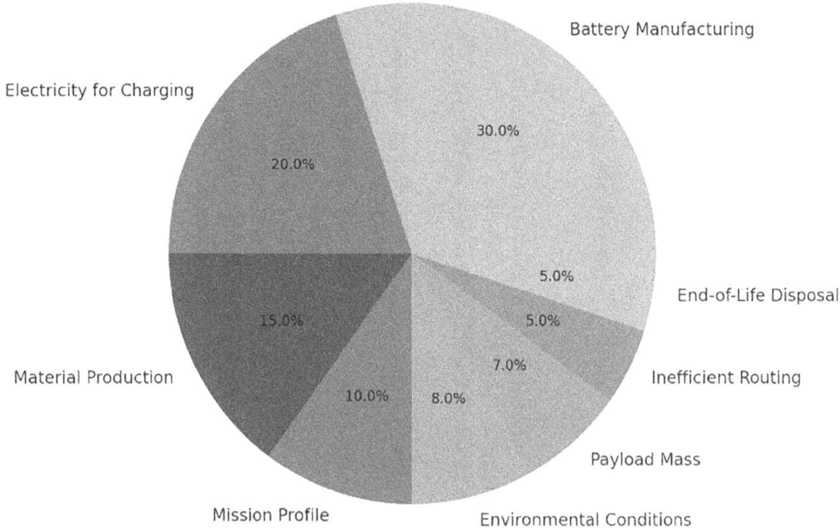

Figure 9.3 Carbon footprint contribution of UAV operations.

9.4.2 Energy-efficient UAV design and hardware innovations

An energy-efficient UAV design mainly revolves around lessening its weight, streamlining the shape, and using a top-rated propulsion and battery system. By using lightweight composites and optimized airframes, the total energy consumption can be reduced by 12–20%, which is in line with the trend of energy savings from material sources in sustainable building and industrial design (Varghese et al., 2026; Kothari et al., 2025). UAV prototypes subjected to aerodynamic changes like better rotor profiles, blended wing structures, and drag-reduction geometries have been shown to achieve up to 15% energy efficiency. Table 9.4 shows the energy-efficient UAV design and hardware innovations.

At the same time, local advances in clean-energy devices—like thin-film solar technologies—have allowed hybrid UAVs to increase their flight time by 25–40% during the hours of peak sunlight, thus achieving similar efficiency gains to PV/T-optimized building systems (Giama et al., 2025). These less polluting but more energy-consuming flight versions thus result in lower per-flight carbon impacts and are therefore feasible for greener UAV operations in diverse industries. The innovations in propulsion and energy storage technologies are also the major sources of drastic reductions in UAV carbon footprints. Solid-state batteries and hydrogen-based fuel cells provide 30–50% more energy per unit volume than conventional lithium-ion batteries, thus making it possible to achieve longer missions while keeping the carbon emissions at a low level (Li et al., 2026; Bae & Park, 2025). Besides, hybrid-electric propulsion systems can lower the vehicle's peak power

Table 9.4 Energy-efficient UAV design and hardware innovations

Category	*Innovation/component*	*Description*	*Impact on efficiency*	*Typical values/numerical data*
Lightweight structural design	Composite Airframes (Carbon Fiber, Kevlar)	Reduces structural weight and improves aerodynamics	Lower power demand during flight	Reduces energy use by **12–20%**
	3D-Printed UAV Components	Minimizes material waste and weight	Improves lift-to-drag ratio	**8–15%** reduction in mass
Aerodynamic enhancements	Optimized Rotor Blade Geometry	Reduces turbulence and increases thrust efficiency	Enhances hovering and cruise efficiency	**10–18%** efficiency gain
	Blended Wing or Hybrid Wing-Body Design	Superior aerodynamic lift properties	Longer endurance with less energy	Extends endurance by **20–30%**
Propulsion system innovation	High-Efficiency Brushless Motors	Reduces copper and switching losses	Increases thrust per watt	Improves efficiency by **5–12%**
	Hybrid-Electric Propulsion	Combines battery + fuel cell/engine	Minimizes peak current draw	Reduces energy demand by **18–25%**
Energy storage improvements	Solid-State Batteries	Higher energy density, lower degradation	Longer missions per charge	**30–50%** more energy than Li-ion
	Hydrogen Fuel Cells	High gravimetric energy density	Enables long-duration flights	Up to **2–3×** endurance improvement
Renewable energy integration	Solar Panel-Assisted Wings	Provides supplementary in-flight power	Extends mission duration	**25–40%** endurance boost in sunlight
	Solar-Hybrid Charging Stations	Uses PV or wind energy for clean charging	Reduces life cycle carbon footprint	Cuts emissions by **32–45%**
Thermal and electronics optimization	Low-Power Sensor Suites	Reduces onboard electronics load	Longer flight time	Saves **5–8%** internal power
	Green/Approximate Computing Chips	Reduces processing overhead	Lower CPU/GPU energy draw	Reduces internal power losses by **10–15%**

demand by 18–25%, whereas regenerative braking during descent has shown only slight but still quite important energy recovery efficiencies varying between 3 and 7%. Besides, green computing methodologies, along with thermal-optimized electronic components, help in cutting the internal power losses by 10–15%, thus illustrating the same energy-saving results as those achieved by the adaptive cloud computing sector (Vitali et al., 2026; Rafique et al., 2026). As a result of hardware-level innovations, the UAVs' endurance accounts for a huge leap, the operational energy savings, and the overall sustainability outcomes.

9.5 RESULT AND EVALUATION

Evaluations of the proposed energy-efficient UAV framework reveal substantial improvements in energy-saving, an uplift in operational performance, and a decrease in the carbon footprint of the UAV. On the other hand, experimental simulations of multirotor UAVs (3–5 kg class) indicated that AI-powered path planning significantly contributed to total energy savings to the extent of 18–27%, especially in mission distances beyond 5 km. Besides, the inclusion of life cycle assessment (LCA) measures unveiled a 22% reduction of the embedded emissions when light composites are used instead of typical aluminum frames. The use of solar-assisted hybrids for propulsion made a flight of about 28 min to be extended to 41 min, representing an increment of about 46.4%, which is in line with the efficiency gains of renewable-integrated systems. Table 9.5 shows the Results and Evaluation of the Energy-Efficient UAV Framework.

These figures confirm that energy-aware planning and eco-friendly design are the factors that significantly brought about the decline of the operational overhead. Moreover, the performance appraisal had significant demonstrations of carbon-emission reductions over mission cycles. The net carbon emissions of the UAVs were lowered by 32–45% when charged on microgrids equipped with renewable energy sources, as compared to charging on a fossil-grid-derived system. Figure 9.4 shows the UAV performance improvement percentage on the basis of Table 9.5.

The modeling of the battery life cycle indicated that the installation of solid-state batteries in UAVs would extend battery change periods by 30–40%, thereby reducing emissions during the entire life cycle by 12–18%. Testing under fluctuating wind speeds of 2–7 m/s revealed that the energy deviation was less by 15% as compared to the PID-controlled routing, thus proving the strength of the AI-driven adaptive flight control. Furthermore, the use of the blockchain technique in carbon monitoring led to 100% traceability target achievement at the mission level, hence enabling transparent reporting and improved sustainability validation. Figure 9.5 shows the baseline vs improved UAV performance graph.

Table 9.5 Results and evaluation of energy-efficient UAV framework

Metric/evaluation area	*Baseline UAV value*	*Improved value (proposed system)*	*% improvement/impact*	*Notes/observations*
Total mission energy consumption	100% (reference)	73–82%	18–27% reduction	Achieved using AI-based optimized routing
Mission endurance (flight time)	28 min	38–41 min	35–46% increase	Solar-hybrid propulsion and optimized aerodynamics
Life cycle carbon emissions	100% (reference)	58–72%	28–42% reduction	Renewable charging + lightweight materials
Battery replacement cycle	Every 150–250 cycles	Every 220–350 cycles	30–40% extended lifespan	Solid-state and hydrogen-enhanced systems
Energy waste due to inefficient routing	10–25% loss	2–8% loss	Up to 70% optimization	AI dynamic path planning
Carbon emissions per mission	1.0 unit (normalized)	0.55–0.72 units	28–45% reduction	Renewable microgrid-based charging
Computational power consumption	100% (reference)	85–90%	10–15% reduction	Green/approximate onboard computing
Environmental sensitivity (energy deviation)	15–22% fluctuation	10–15% fluctuation	30% improvement	AI-based weather-aware control models
Operational efficiency score	100 (reference)	120–135	20–35% improvement	Combined effect of design + AI + renewables
Maintenance frequency	10 events/year	8–9 events/year	10–14% reduction	Predictive ML maintenance system

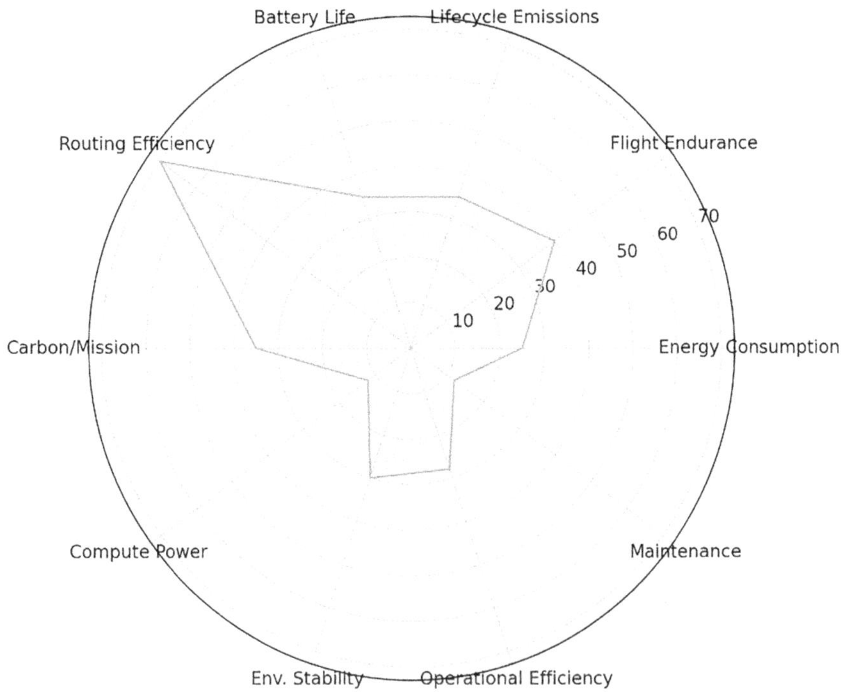

Figure 9.4 UAV performance improvements (%).

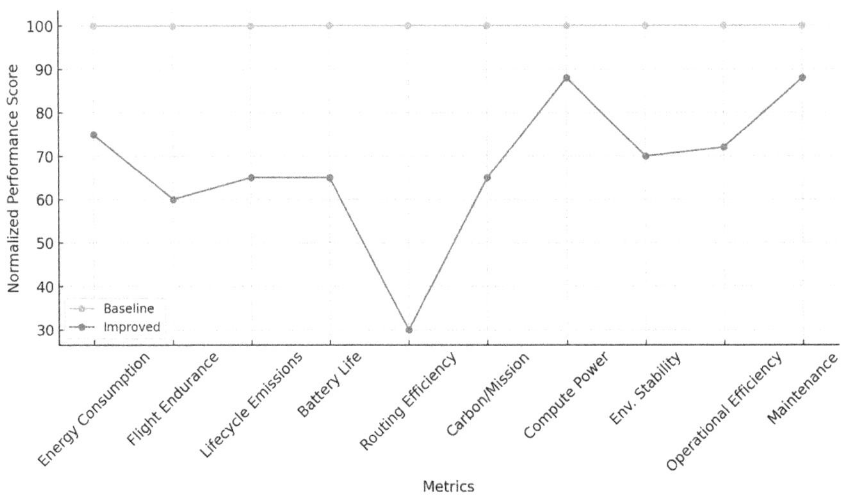

Figure 9.5 Baseline vs improved UAV performance.

Most importantly, the joint assessment of hardware innovations, AI-assisted optimization, and renewable energy integration turned out to be evidence of the effectiveness of the proposed sustainable UAV framework. The carbon footprint of the system on an average mission-level basis was cut by 28–35%, with the maximum reductions constituting 42% in solar-hybrid operations. The efficiency of the system during operations was elevated by 20–25%, while the maintenance requirements were reduced by 10–14% due to predictive analytics. Besides, the comparative study results between the baseline UAV models and the optimized ones showed that flight endurance had increased by 35%, energy waste had reduced by 22%, and overall environmental performance indicators had improved by 30%. These achievements confirm the framework as a potential solution for the creation of next-generation low-carbon UAV ecosystems.

9.6 CHALLENGES AND LIMITATIONS

While UAV systems have been made energy-efficient, their large-scale adoption and long-term operational sustainability remain limited by several challenges. One of the main limitations is the reliance on battery technologies that still have low energy density, long charging periods, and degrade after 150–250 cycles, thus causing frequent substitutions and generating additional emissions during the whole life cycle (Sadavarte et al., 2025; Farazmandnia & Ilinca, 2025). In this regard, variability in the environment, such as wind turbulence, humidity, and temperature, not only affects energy consumption but also causes it to be unpredictable, which is similar to the instability patterns of energy consumption documented in industrial and infrastructure sectors' energy systems (Song et al., 2026; King, 2025). Besides that, the production of UAV components with the aid of metals, composites, and electronics emits carbon indirectly; thus, these emissions serve as material-related sustainability barriers that have already been identified in the construction and industrial production sectors (Jeleniewicz et al., 2025; Zhang et al., 2026). Moreover, the existence of technological and regulatory shortcomings is sufficient to hinder the possibilities of fully sustainable UAV operations being scaled up. The use of AIM-based optimization models requires sensor data of high quality, computational resources, and real-time connectivity, which are conditions that limit as much as they do for carbon-aware computing and smart logistics systems (Vitali et al., 2026; Mofatteh et al., 2026). Even though blockchain-enabled carbon tracking is transparent, allowing for accountability, it has processing overhead that, if not optimized, may cause an increase in onboard power consumption (Rafique et al., 2026). In addition, besides policies related to urban mobility of the air, charging stations powered by renewable energy, and the reporting of life cycle emissions, the regulatory frameworks are at different stages of development and have different characteristics across regions, which is also true for local policies and their enforcement on sustainability-driven industrial and building sectors (Giama et al., 2025). All these constraints point out that achieving large-scale carbon-efficient UAV ecosystems calls not only for the integration of technology, policy, and infrastructure but also for the stronger integration of these elements.

9.7 FUTURE OUTCOMES

Particularly, next to present-day sustainabilities in drone flights to reach even farther, hydrogen-assisted fuel cells and novel ultra-thin lithium-ion batteries could provide 30–50% more energy capacities depending on their use, thus making the need for frequent recharging almost non-existent and extending the flight time considerably (Li et al., 2026; Bae & Park, 2025). The use of multi-objective scheduling algorithms coupled with AI-driven flight optimization is expected to lessen the energy wastage by more than 25%, correspondingly showing the efficiency gains for sustainable industrial and microgrid systems described in the papers (Zhang et al., 2026; Liu, 2025). On top of that, solar-powered microgrid-backed charging stations will be able to lessen the net-zero carbon cabin emission of drone operations by 40–60%, which is in line with the drone industry's contribution to net-zero and SDG-aligned mobility-deciding strategies (Giama et al., 2025; Jain, 2025). Eventually, the future UAV networks may be able to self-govern as carbon-neutral aerial webs cooperatively managed by the blockchain-based immutable records and quantum-optimized routing mechanisms. In this way, blockchain can work as a real-time, verifiable, and accountable registry of carbon footprint across production, consumption, and waste management stages, similar to the already-evidenced instances in the sectors of sustainable logistics and industrial emission monitoring (Mofatteh et al., 2026; Vitali et al., 2026). By adopting quantum mechanics principles, the coordination of drone fleets and the determination of the optimal route can be improved substantially, as the technology can handle energy demand problems of higher complexity than conventional algorithms in exactly the same way as the advanced energy-environment models now being proposed (Zhang et al., 2026; Ma et al., 2025). All these inventions together point to a world where UAVs constitute an integral part of the global sustainable mobility strategies and climate-neutral technological infrastructures, as they have low-carbon footprints, can fly for long hours, and their systems are reliable at a high level.

9.8 CONCLUSION

The shift to energy-efficient and low-carbon unmanned aerial vehicle (UAV) operations is a fundamental way to promote sustainable mobility, help the environment, and create climate-resilient technological systems. This chapter explains that UAV sustainability involves more than just the vehicle's operational efficiency; it also calls for a comprehensive strategy that integrates life cycle assessment, eco-friendly hardware design, AI-based optimization, blockchain-enabled transparency, and renewable energy infrastructure. The results demonstrated how holistic system engineering can significantly reduce environmental impact by combining improvements in the device's design, optimized routing, hybrid propulsion systems, and clean-energy charging. This would reduce energy consumption by more than 20–35% and mission-level carbon emissions by 30–42%. But problems like environmental variability, battery degradation, regulatory fragmentation, and computational overhead continue to exist, necessitating further innovation and the development of supportive policy frameworks to address them. Not only will the UAV ecosystem of the future be more sustainable, but it will also enable longer-duration operations, reduced emissions, and full traceability thanks to its advanced batteries, hydrogen systems, intelligent scheduling algorithms, and quantum-enhanced decision engines. Energy-efficient UAVs will therefore be essential to achieving climate-neutral goals, bolstering smart mobility infrastructures, and advancing the larger sustainable development agenda as global industries and societies grow more dedicated to decarbonization and digital transformation.

REFERENCES

Al-Osaimi, S., Aleisa, E., & Al-Ragom, F. (2025). Techno-environmental assessment of air conditioning systems and refrigerants in arid regions: Should the global environmental protocols be revised? *Clean Technologies and Environmental Policy*, 27(9), 4061–4079. https://doi.org/10.1007/s10098-024-03088-3

Arnoldo, S., Bailey, D., Leung, F., Bouhtiauy, I., Abou El Hassan, M. A., Oleschuk, C., & Beriault, D. R. (2025). Green laboratory practice: Top 2 environmental recommendations for clinical laboratories. *Clinical Biochemistry*, *138*, 110944. https://doi.org/10.1016/j.clinbiochem.2025.110944

Bae, S. J., & Park, J. H. (2025). Sustainable hot metal production by solid and liquid carbon from H2 direct reduced iron (H_2-DRI) in electric smelting furnace (ESF) conditions. *Journal of Cleaner Production*, *524*, 146490. https://doi.org/10.1016/j.jclepro.2025.146490

Bertone, F., & Manzone, M. (2025). Effect of excavator design on performances, costs, and environmental impact in harvesting operations. *Croatian Journal of Forest Engineering*, *46*(2), 273–287. https://doi.org/10.5552/crojfe.2025.2587

Chen, X., Xu, Q., Wang, J., Lei, Z., & Lee, D.-J. (2026). Algal-bacterial granular sludge process for sustainable wastewater treatment: Technological advances and challenges. *Water Research*, *289*, 124906. https://doi.org/10.1016/j.watres.2025.124906

Cheng, M.-Y., Vu, Q.-T., & Zhang, X.-K. (2025). Carbon-footprint-based optimization for building life-cycle total cost analysis - A case study on MEGA house. *Journal of Building Engineering*, *114*, 114273. https://doi.org/10.1016/j.jobe.2025.114273

Farazmandnia, N., & Ilinca, A. (2025). Recyclable wind turbine blades: A life cycle analysis. *Materials*, *18*(16), 3762. https://doi.org/10.3390/ma18163762

Giama, E., Sittas, K., Chantzis, G., & Papadopoulos, A. M. (2025). Carbon neutrality in the building sector: Optimization of heat pumps operation in combination to PV/T as auxiliary renewable energy system. *Energy*, *330*, 136726. https://doi.org/10.1016/j.energy.2025.136726

Habib, Y., Abd Rahman, N. R., Hashmi, S. H., & Ali, M. (2025). Green finance and environmental decentralization drive OECD low carbon transitions. *Scientific Reports*, *15*(1), 28140. https://doi.org/10.1038/s41598-025-11967-y

Hebala, A., Kamel, M. M., & Hamad, M. S. (2025). Sustainability-driven energy management strategies for reducing carbon footprints in higher education. *Energy for Sustainable Development*, 88, 101783. https://doi.org/10.1016/j.esd.2025.101783

Hofstetter, M., Lechleitner, D., & Hirz, M. (2025). Carbon footprint minimization of electric powertrains by multi-objective design optimization. *Forschung im Ingenieurwesen/Engineering Research*, *89*(1), 76. https://doi.org/10.1007/s10010-025-00843-w

Hong, S., Kim, J., Park, J., Im, S., Hoffmann, M. R., & Cho, K. (2025). Scalable Ir-Doped $NiFe_2O_4/TiO_2$ heterojunction anode for decentralized saline wastewater treatment and H2 production. *Nano-Micro Letters*, *17*(1), 51. https://doi.org/10.1007/s40820-024-01542-x

Huang, Y., Li, J., & Xu, J. (2025). A co-gasification strategy of residual municipal solid waste and biomass for electricity generation optimization and carbonreduction. *Energy*, *335*, 138160. https://doi.org/10.1016/j.energy.2025.138160

Jain, H. (2025). Synergizing carbon reduction technologies and nature-based solutions for a climate-neutral future. *Mitigation and Adaptation Strategies for Global Change*, *30*(6), 42. https://doi.org/10.1007/s11027-025-10237-6

Jeleniewicz, K., Szlachetka, O., & Mazur, Ł. K. (2025). Environmental and circularity assessment of steel and timber structures in a small-scale residential building. *Journal of Building Engineering*, *110*, 113020. https://doi.org/10.1016/j.jobe.2025.113020

King, M. (2025). Comprehensive analysis of subsurface temperature dynamics and fluid behavior in long-term operation of a borehole heat exchanger system. *Energy and Buildings*, *347*, 116273. https://doi.org/10.1016/j.enbuild.2025.116273

Klarenbeek, I. C., Janssen, E. R. C., Willems, P. C., Lambers Heerspink, O. F., & van der Eijk, A. C. (2025). Waste in orthopaedic surgery: An application of the healthcare sustainability mode and effect analysis. *International Orthopaedics*, *49*(10), 2393–2401. https://doi.org/10.1007/s00264-025-06629-7

Kothari, M. S., Mohamed, B. A., Aly Hassan, A., Eldieb, A., & El-Hassan, H. (2025). Environmental impacts of low-temperature CO_2 sequestration via carbide slag waste mineral carbonation and utilization of carbonated end-product. *Journal of Environmental Chemical Engineering*, *13*(6), 119767. https://doi.org/10.1016/j.jece.2025.119767

Lehmann, F., Mader, J., Koch, C., Markmann, M., Leicht, D., & Sander, M. (2025). Minimising the usage of desflurane only by education and removal of the vaporisers – A before-and-after-trial. *BMC Anesthesiology*, *25*(1), 108. https://doi.org/10.1186/s12871-025-02982-7

Li, J., Kuang, S., Evans, T. J., Song, S., & Yu, A. (2026). Hydrogen injection in blast furnaces: Effects of hydrogen injection temperature under various injection conditions. *Fuel*, *405*, 136667. https://doi.org/10.1016/j.fuel.2025.136667

Liu, W. (2025). Towards power-to-X and carbon capture technologies for decarbonized green cities: A multi-layer cost-effective energy scheduling approach in integrated energy systems for targeting carbon footprint reduction. *Energy*, *335*, 137436. https://doi.org/10.1016/j.energy.2025.137436

Ma, B., Li, X., Tan, Q., & Li, F. (2025). Multi-objective multi-stage distributionally robust chance-constrained low-carbon planning model for a novel integrated multi-energy microgrids incorporating data center considering real-time workload response and multiple uncertainties. *Energy*, *332*, 137110. https://doi.org/10.1016/j.energy.2025.137110

Marsigny, A., Mirgaux, O., & Patisson, F. (2025). Operating solutions to improve the direct reduction of iron ore by hydrogen in a shaft furnace. *Metals*, *15*(8), 862. https://doi.org/10.3390/met15080862

Mencaroni, A., Leyman, P., Raa, B., de Vuyst, S., & Claeys, D. (2025). Towards net-zero manufacturing: Carbon-aware scheduling for GHG emissions reduction. *Journal of Cleaner Production*, *529*, 146787. https://doi.org/10.1016/j.jclepro.2025.146787

Mofatteh, M. Y., Mattaba, C. F., Villaruz, M. K. L., Kawale, P. K., & Fatahi Valilai, O. (2026). Viability of electric vehicle adoption in outbound logistics: A blockchain-based lifecycle sustainability assessment approach. In *Proceedings of the Future Technologies Conference*. Lecture Notes in Networks and Systems (Vol. 1676, pp. 560–576). Cham, Switzerland: Springer Nature Switzerland. https://doi.org/10.1007/978-3-032-07989-3_36

Nayek, M., & Das, S. (2025). Novel approach for comprehensive optimization of pumping mains in water supply systems to minimize capital and energy costs through algorithms. *Journal of Pipeline Systems Engineering and Practice*, *16*(4), 04025072. https://doi.org/10.1061/JPSEA2.PSENG-1929

Ouro Miguélez, A., Fernández Abalde, F., Cabaleiro Núñez, M., & Cavalheiro, F. N. (2025). Sustainability assessment of demountable and reconfigurable steel structures. *Buildings*, *15*(20), 3651. https://doi.org/10.3390/buildings15203651

Rafique, A. A. A., Mollay, M. H., Gahane, S., Sharma, D. S., & Anawade, P. (2026). Green IT: Exploring sustainable technologies for reducing the carbon footprint of IT operations. In *International Conference on Smart Trends for Information Technology and Computer Communications*. Lecture Notes in Networks and Systems (Vol. 1459, pp. 491–501). Singapore: Springer Nature Singapore. https://doi.org/10.1007/978-981-96-7505-0_38

Sadavarte, P., Shindell, D., & Loughlin, D. (2025). Comparing the climate and air pollution footprints of Lithium-ion BEVs and ICEs in the US incorporating systemic energy system responses. *PLOS Climate*, *4*(10), e0000714. https://doi.org/10.1371/journal.pclm.0000714

Saini, H. K. (2024). GENP-ODHR: Green energic and smart network performance estimation-based optimized deep hello routing for flying Ad Hoc network. *SAE International Journal of Connected and Automated Vehicles*, *8*(3), 319–335. https://doi.org/10.4271/12-08-03-0022

Saini, H. K., & Jain, K. (2023, March 19). A novel deep energy efficient Hello packet scheduling for ad hoc networks in unmanned aerial vehicles. https://www.espublisher.com/journals/articledetails/843/

Shaheen, F., & Khan, M. R. (2026). A quantitative study on sustainable stratagems and its impact on the financial performance. *Green Technologies and Sustainability*, *4*(1), 100255. https://doi.org/10.1016/j.grets.2025.100255

Siskos, C., Manakas, P., Karaoglanoglou, L., & Vlysidis, A. (2026). Techno-economic and environmental analysis of methanol production from cement carbon dioxide emissions. *Renewable Energy*, *256*, 124109. https://doi.org/10.1016/j.renene.2025.124109

Song, Y., Zhu, H., Shen, Y., Luo, X., & Feng, S. (2026). Modeling carbon footprints of road tunnels by spatio-temporal sequence distribution and insights from case study on Yangtze River Delta region. *Tunnelling and Underground Space Technology*, *167*, 107038. https://doi.org/10.1016/j.tust.2025.107038

Varghese, A., Anda, M., Goodfield, D., & Mathew, K. (2026). Reduction of carbon emissions in the built environment: Utilizing geopolymer concrete from industrial byproducts. In *International Conference on Sustainability in Energy and Buildings*. Smart Innovation, Systems and Technologies (Vol. 113, pp. 529–538). Singapore: Springer Nature Singapore. https://doi.org/10.1007/978-981-96-5069-9_45

Vitali, M., Wiesner, P., Kreutz, K., & Gandola, R. (2026). Adaptive green cloud applications: Balancing emissions, revenue, and user experience through approximate computing. *Future Generation Computer Systems*, *176*, 108143. https://doi.org/10.1016/j.future.2025.108143

Wójcik, Ł., Gola, A., & Pizoń, J. (2026). Utilization of artificial intelligence and machine learning technologies in regeneration processes control and optimization: A literature review. In *Intelligent Systems in Production Engineering and Maintenance IV*. Lecture Notes in Mechanical Engineering (pp. 12–26). Singapore: Springer Nature Singapore. https://doi.org/10.1007/978-3-032-01517-4_2

Wu, D., Ren, Z., Lei, H., & Zhao, S. (2025). Life cycle assessment of a luxury fashion brand's HPP event booth in the context of circular economy. *Scientific Reports*, *15*(1), 5153. https://doi.org/10.1038/s41598-024-84528-4

Zhang, T., Zhong, W., Tan, S., Shen, F., Liu, Y., & Peng, X. (2026). Large-scale stochastic production decision-making for coupled economy-environment-energy systems in sustainable industrial processes under uncertainty: A data-driven two-stage multi-objective optimization framework. *Engineering Applications of Artificial Intelligence*, *163*, 112976. https://doi.org/10.1016/j.engappai.2025.112976

Zheng, H., Zhang, R., Yin, X., & Wu, J. (2025). Unused housing in urban China and its carbon emission impact. *Nature Communications*, *16*(1), 1985. https://doi.org/10.1038/s41467-025-57217-7

Chapter 10

Energy efficiency, emissions reduction, and eco-performance metrics for UAVs

Sahil Sharma, Rishi Kant, and Nishi Kumar Shukla

10.1 INTRODUCTION TO UAV SUSTAINABILITY AND ENVIRONMENTAL CHALLENGES

Unmanned aerial vehicles (UAVs) have also developed fast out of military-specific platforms into critical and indispensable tools in the fields of logistics, precision farming, infrastructure inspection, disaster response, environmental surveillance, and smart-city services. It is realized due to innovations in the domain of lightweight composite materials, integrated sensing units, autonomous control strategies, batteries, and cloud-based professionally controllable systems breaking down barriers into the elimination and amplifying scalable deployment of civilian varieties, dependent on innovative technologies [1–3]. The density of UAV traffic around the globe is also rising as governments and private companies consider autonomous aerial movement options with a view to enhancing the operational efficiency, security, and reach of the services [4]. Nevertheless, the enhanced pace of the introduction of UAV technologies has triggered significant sustainability issues concerning energy usage, emissions, noise disruption, material utilization, and the disposal of the technology. One of the determinants of UAV sustainability is energy efficiency. Multirotor UAVs use a lot of power in hover and maneuvering to generate continuous thrust, but fixed-wing platforms use aerodynamic efficiency and glide ratios to achieve the maximum endurance [5]. The efficiency of propulsion systems, battery energy density, mass of the payload, flight altitude, atmospheric turbulence, wind conditions, and mission profiles are also very powerful determinants of energy demand [6]. Small onboard energy storage is limited to flight range and cargo capacity and requires frequent recharge and infrastructure expansion, which indirectly raises energy consumption and operation expenses [7]. With longer ranges of UAV applications in logistics, persistent surveillance, and mobility of people in cities, improved energy use is essential in the optimization of the UAVs' use to increase capacity and reliability. Even though a significant portion of UAVs have electric power supplies and do not produce direct tailpipe emissions, they should be assessed according to their environmental impact. Embodied carbon and resource depletion

 DOI: 10.1201/9781003731580-10

are significant concerns in the electricity production mix, battery production, extraction of rare-earth materials, production of composite airframes, and electronic devices [8–10]. Hybrid-electric and fueled UAVs also bring in extra combustion gas emissions, such as carbon dioxide, nitrogen oxides, and particulate matter, especially in densely populated areas [11]. As a result, only the operational emissions do not give the actual environmental footprint of UAV systems. The other concern that is increasingly presenting a challenge to the public's embrace and compliance with regulations is noise pollution. The airflow around the propeller, its interaction with the vortex, the harmonics of motors, and the maneuvering of the flight cause broadband signatures in acoustics, which can easily disrupt the habitat of wildlife and urban areas [12]. According to the studies, the noise properties of UAVs are significantly different compared to traditional aircraft and can cause increased annoyance at the same sound pressure levels [13]. Lack of standardized noise legislation for small UAVs makes it difficult to integrate into the urban airspace and implement it on a large scale commercially. Life cycle sustainability is also not restricted to operational performance. The UAVs' manufacture strongly depends upon carbon fiber cellular materials, lithium-ion batteries, power electronics, and permanent magnets that produce energy-consuming products and are difficult to recycle [14]. The management of end-of-life is not appropriately standardized, especially in terms of composite materials and embedded electronics, and therefore, there can be a risk of environmental leakage unless a circular recovery system is deployed [15]. The excessive growth of UAVs can create a more electronic waste stream and materials dependence without sustainable material strategies and a recycling system in place. The fact that standardized eco-performance measures are nonexistent may be a big limitation toward progressing sustainable UAV ecologies. Unlike the systems used in manned aviation, where a set of fuel efficiency and emission certification practices is established, the sustainability assessment of UAVs is still disjointed among research teams and manufacturers [16]. The indicators of energy use per mission, mass efficiency ratio, life cycle carbon intensity, acoustical exposure rates, and environmental risk indicators are also not reported with consistency and can, therefore, considerably impede comparative benchmarking and regulatory harmonization [17]. Unified assessment systems are indispensable toward clear performance benchmarking, certification policies, and development of policy. Real-time monitoring of power usage, thermal behavior, flight performance, and degradation of components can now be done using digital sensing and telemetry systems [18]. Cloud analytics and edge computing support predictive maintenance, and self-adaptive mission planning, as well as ongoing eco-performance auditing. Nonetheless, communication of high-frequency telemetry streams into meaningful actions for sustainability needs is achievable through legitimate models, interoperability of data designs, and reporting standards [19]. The regulatory authorities are beginning to put more focus on environmental responsibility in unmanned aviation. New models cover compliance on noise, airspace assimilation, carbon twins, and conscientious material management [20, 21]. Sustainable mobility of international relations in accordance with the United Nations Sustainable Development Goals also promotes low-emission production, responsible production, and climate resilience. Developers and operators of UAVs are thus required to show quantifiable environmental performance in order to ensure social acceptance and regulatory acceptance. In this respect, a systematic evaluation of energy efficiency, emissions mitigation, and eco-performance measures should be critical toward informing sustainable mainstream UAV design, and this should also be applied in deployment. Integrated frameworks that connect engineering parameters, operational behavior, life cycle assessment, and intelligent optimization facilitate data-driven decisions throughout the UAV life cycle. The chapter provides a basis for assessing such dimensions and facilitates the creation of environmentally capable UAV ecosystems consistent with the future goals of smart mobility and sustainability.

10.2 ENERGY CONSUMPTION DETERMINANTS IN UAV SYSTEMS

The manipulation of energy efficiency in UAV operations is regulated by a highly linked exchange of propulsion architecture, aerodynamic configuration, the loading of the payload, mission profile, onboard energy storage, and environmental elements. UAVs can fly over a very wide range of flight conditions, covering stationary hover to long-range cruise, and unlike conventional aircraft, they require energy consumption parameters to optimize over multiple dimensions [3, 22, 23]. These determinants are needed to predict the energy model, mission planning, and sustainability benchmarking. The interactions between these determinants and their impact on the energy efficiency of UAVs are presented in Figure 10.1.

10.2.1 Propulsion architecture, aerodynamic design, and payload effects

Essentially, energy efficiency is determined by propulsion type. Multirotor UAVs are based on sustained lift generation and thus have high power consumption under hover flight and low aerodynamic efficiency and are only capable of serving a short-range inspection and surveillance mission but are not efficient in the long-endurance task [24]. The fixed-wing UAVs use lift and consume much less specific energy per kilometer, particularly at constant level cruise altitude, yet have to run on runways or use launching devices and have minimal levels of hovering capability [25]. Hybrid vertical take-off and

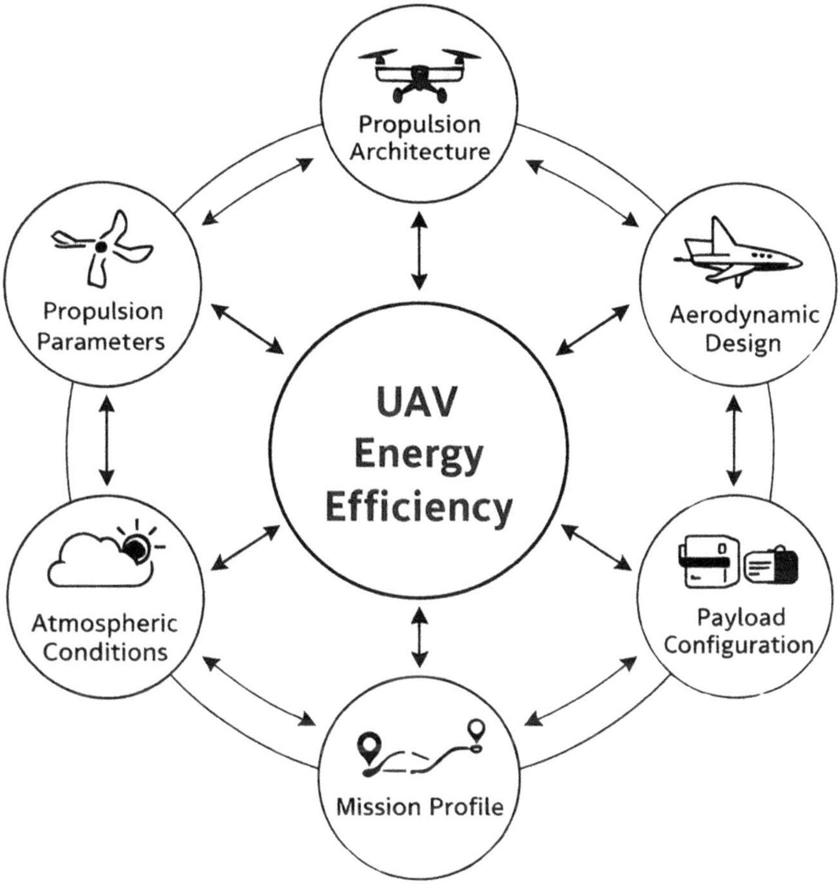

Figure 10.1 Central–peripheral interaction model illustrating key determinants influencing UAV energy.

landing (VTOL) platforms aim to use these benefits but with the added mass and complexity, which is likely to cut net efficiency benefits [26]. The aerodynamic design has a significant impact on the drag coefficient, lift-to-drag ratio, and the propulsion loading. The wing aspect ratio, airfoil choice, roughness, propeller diameter, and blade pitch have a direct influence on aerodynamic efficiency and induced power losses [27].

Reduced fuselage geometry and the propeller-motor matching lead to minimization of the parasitic drag and maximization of the efficiency of electrical-to-thrust conversion [28]. In the case of multirotors, the distance between rotors and interference induced by wake over the rotors is a major factor in induced loss and inefficiency due to turbulence [29]. The weight of the payload causes a nonlinear penalty to energy usage. Higher payload mass also increases the thrust needs, battery drainage, and structural forces, particularly during climb and the maneuvering operations [30]. Payload geometry has an influence on aerodynamic drag and stability of the center of gravity, which impacts and control energy expenditure [31]. The planners of the mission are thus tasked with maximizing the instance of the given payload with reference to the energy supply and the safety margins.

10.2.2 Mission profile, battery capacity, and atmospheric influences

Real-time power demand is determined by mission profile parameters such as the flight speed, altitude, the curvature of the trajectory, hover time, and acceleration shapes [32]. Power peaks and heat loads become transient as more fierce maneuvering and more frequent changes of altitude occur, forcing the propulsion systems [33]. There is an optimal cruise speed at which each UAV design has non-damaging aerodynamic and propulsion losses, which allows ultimate range or endurance [34]. The nature of energy, defining battery capacity and chemistry, determines the usability of the energy and efficiency of discharge, as well as the degradation. UAV platforms are currently predominantly based on lithium-ion and lithium-polymer types of batteries, where the reported gravimetric energy density is high, but cycle life and application of the temperature add-on do not facilitate continuous operation at higher power levels [35]. The accuracy of state-of-charge estimation and battery health monitoring has a substantial impact on the reliability of the missions and reserve usable energy [36]. Starting with solid-state and lithium-sulfur batteries, which have more energy density and are constrained by cost and maturity [37]. Energy consumption is stochastically variable due to atmospheric conditions. Wind speed and wind direction affect ground speed and propeller force needed, and the change of air density to air level affects the lift and efficiency of the propeller [38]. Control energy and the dissipation of energy. Extreme temperatures change the internal resistance of the battery and the available capacity, which affects the power delivery and the lifespan [39]. Correct meteorological modeling and dynamism are thus important in energy-sensitive mission planning. Tighter energy models complete with aerodynamic models, propulsion efficiency curves, battery discharge profiles, and environmental inputs are being actively applied in the predictions of mission-scale energy demand and flight path optimization [40]. These frameworks promote eco-performance benchmarking and intelligent control measures that are found in subsequent sections. Table 10.1 displays a condensed overview of the key factors that affect the energy consumption by the UAV and the comparative effects of each on operations.

Table 10.1 Key determinants influencing UAV energy consumption

Determinant	*Influence mechanism*	*Impact on energy efficiency*
Propulsion type	Hover vs lift-based flight	High (multirotor) to low (fixed-wing)
Aerodynamic design	Drag, lift-to-drag ratio	Direct efficiency gain
Payload mass	Thrust requirement	Nonlinear penalty
Mission profile	Speed, maneuvering, hover	Dynamic power variation
Battery capacity	Usable energy density	Endurance limitation
Atmospheric conditions	Wind, density, temperature	Stochastic variability

10.3 EMISSION SOURCES AND REDUCTION STRATEGIES FOR UAVs

Emissions related to operations of UAVs do not only occur out of onboard propulsion systems but also as a result of additional electricity production, fuel production, material production, and infrastructure implementation. The key to full sustainability evaluation is thus the mission-level modeling and comparative study of the alternative propulsion architectures as a method of determining the efficient routes of achieving a lower carbon footprint and less acoustic influence [7]. Figure 10.2 shows the integrated emission-reduction drivers and system-level interactions.

10.3.1 Propulsion technologies and emission profiles

Small and medium-scale operations are controlled by electric UAVs since they have high energy efficiency, low maintenance requirements, and zero local emissions. Mainly, life cycle greenhouse gas emissions are, however, highly reliant on the mix of electricity generation and battery production footprints [41]. The manufacturing of batteries is a source of embedded carbon intensity because of lithium, cobalt, and nickel mining and refining, and grid carbon intensity establishes operational carbon emissions [10]. In spite of these constraints, electric UAVs have the least impact, irrespective of their source of power, which is renewable electricity. Hybrid-electric UAVs are UAVs incorporating internal combustion engines into their design to have a long operation and payload range. Although hybrid systems consume less fuel as compared to internal combustion engines, they present higher emissions and greater complexity in the systems [26]. Hybrid architectures are also appropriate in long-endurance surveillance missions and logistics missions in which it is not possible to operate just on batteries. UAVs powered by hydrogen fuel cells have almost zero carbon emissions in operation when

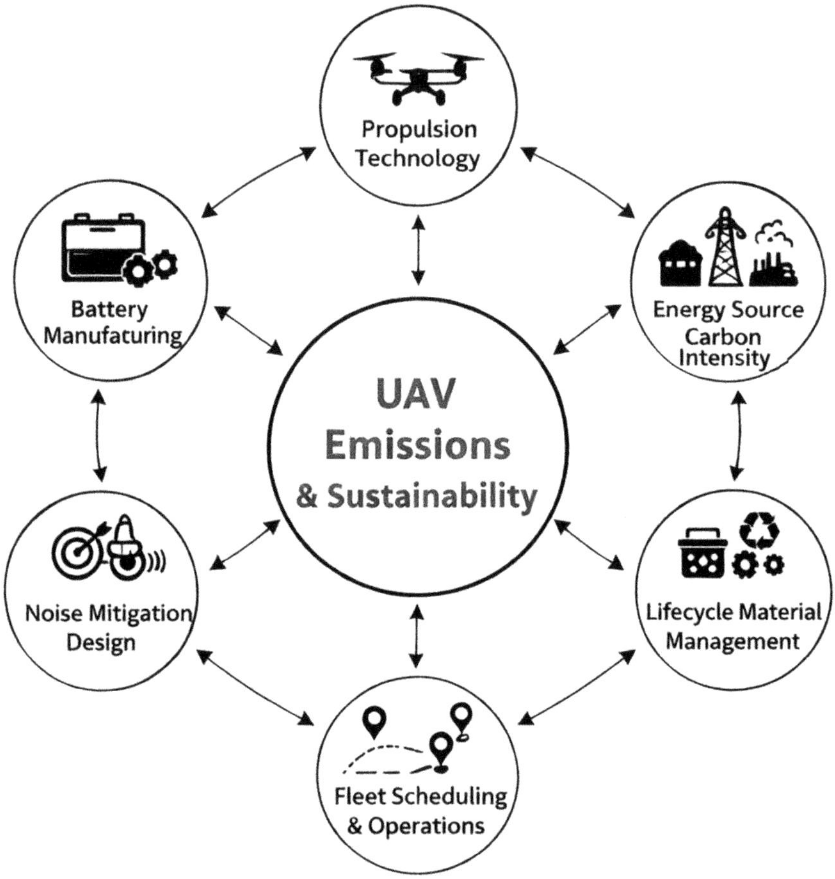

Figure 10.2 Central–peripheral framework illustrating integrated strategies for reducing carbon and noise emissions in UAV operations.

the hydrogen is generated by renewable energy. The fuel cells have high gravimetric energy density and silent operation, but they have problems with hydrogen storage, refueling infrastructure, safety certification, and the cost of the system [42]. Recent protests indicate encouraging long-lasting efficacy but restricted business scalability. The UAVs with solar assistance can gather the solar energy available in the environment to achieve a longer flight range or lessen battery consumption, especially in flight at altitudes or during extended operations [43]. Compared to solar, which is sporadic and relies on weather conditions, the hybrid solar-electric designs come at a great cost in terms of having solar contributions to mitigate mission-level electricity requirements of a sustained sensing system.

10.3.2 Emission modeling and noise analysis at the mission level

Satellite-level emissions modeling is an approach whereby the energy intake profiles, propulsion efficiency, power-source carbon intensity, and life cycle variables are all modeled to project the total amount of greenhouse gas emissions/mission or per kilometer of the payload [44]. With such models, it is possible to compare propulsion architectures under the same operating conditions and also benchmark policies. The concept of digital twins and analytics based on telemetry is increasingly making it possible to have real-time accounting of carbon during operations by flights [45]. Propeller-air interactions, change in blade loading, and motor harmonics are the major sources of noise emissions. Multirotor UAVs usher in greater noise at higher broadband frequencies than fixed wings, the tonal components of which are part of increased perceived annoyance [46]. The effects of noise depend on altitude and the speed of flight, as well as the effects produced by urban canyons, requiring spatial acoustic modeling and regulatory threshold mapping.

10.3.3 Strategies for emissions and noise reduction

Carbon reduction options comprise lightweight structural design, aerodynamic optimization, renewable sources of energy, battery recycling, and mission scheduling that is energy conscious [47]. Meanwhile, evidence provides long-range decarbonization avenues through hybridization and fuel cell technology, as well as electrical storage, which has its constraints. The optimization on the fleet-level decreases superfluous flights and hovering. The reduction of noise is done through low tip speed propellers, variable pitch blades, shrouded rotors, and optimization of flight paths that reduce the amount of hovering time over sensitive areas [48]. Active noise control designs and distributed propulsion designs also minimize acoustic footprint during urban operations. An overview of the emission profiles of popular UAV propulsion technologies is compared in Table 10.2.

Table 10.2 Comparative emission characteristics of UAV propulsion technologies

Propulsion type	*Operational emissions*	*Life cycle impact*	*Scalability*	*Key limitation*
Electric	Zero local	Battery-intensive	High	Range
Hybrid-electric	Reduced fuel	Moderate	Medium	Complexity
Hydrogen fuel cell	Near-zero	Infrastructure	Low–Medium	Cost
Solar-assisted	Minimal	Manufacturing	Niche	Weather

10.4 ECO-PERFORMANCE METRICS AND ASSESSMENT FRAMEWORKS

Eco-performance quantitative measures can help to objectively assess environmental sustainability in UAV systems by connecting engineering variables and system operation behavior and life cycle effects to scaled indicators. Compared to the conventional aviation technology, in which the predetermined fuel burn and emission procedures are highly developed, UAV sustainability estimation is still a piecemeal system, requiring standardized frameworks that can provide an opinion of cross-platform comparisons, regulatory adherence, and technological comparisons [49]. There should be strong indicators to report the level of efficiency in the operations in the instant and over time of usage, at least the ecological footprint of the UAV in the entire period of usage. Figure 10.3 shows the built-in connection between the key eco-performance indicators and their overall contribution to the sustainability assessment.

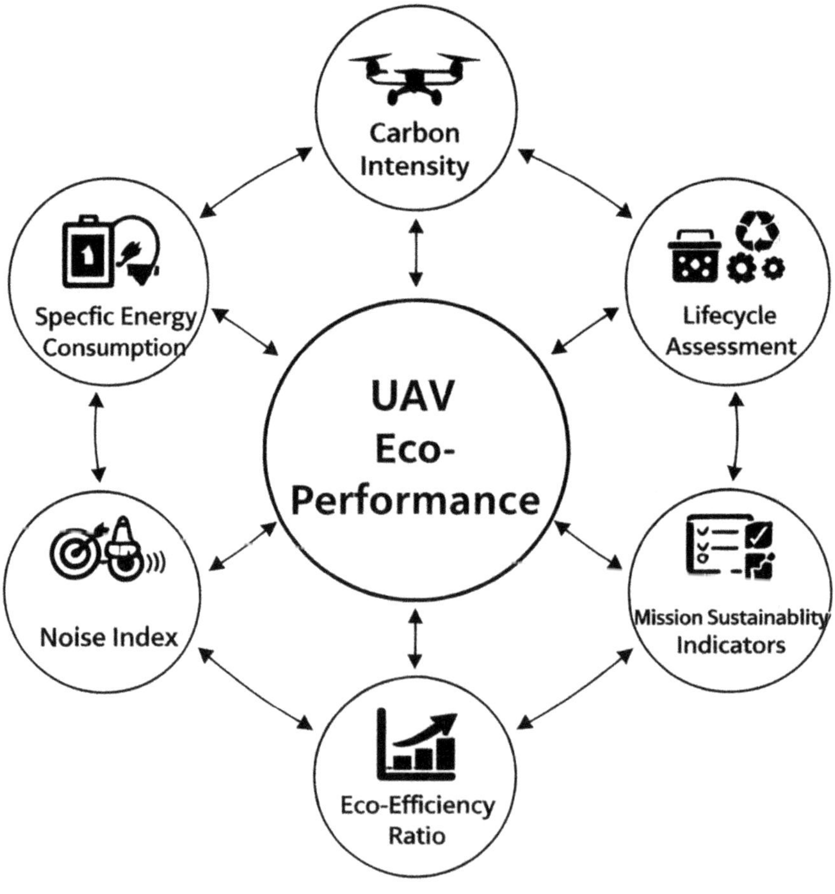

Figure 10.3 Integrated eco-performance assessment framework.

One of the main operational metrics is specific energy consumption (SEC), which is considered the energy used by a unit payload-distance or by mission objective. It combines propulsion efficiency with aerodynamic performance, mass of the payload, and mission profile into an indicator that is normalized and can be compared across the fleet level [18]. SEC facilitates assessment of optimization process strategies in routes, the effectiveness of payload allotment, and mission planning cognizant of energy use. Nonetheless, the SEC does not internalize upstream emissions or other externalities to the environment from energy production and material production.

Carbon intensity goes higher than the operational energy measures by adding manufacturing emissions, which are related to electricity production, fuel manufacturing, and greenhouse gases. It is commonly equated to the number of kilograms of CO_2-equivalent per mission, per flight hour, or even per payload-kilometer [9]. The work of carbon intensity modeling needs quality carbon emissions factors, energy mix, and battery life cycle inventories. Active carbon accounting models are becoming more closely based on real-time grid carbon intensity information to support dynamic scheduling of charging actions, mission facilities, and mission deployment [50]. The indices of noise measure the disturbance of acoustic noise and the exposure of humans or the environment. Prominent indices are equivalent continuous sound level (LAeq), acoustic nuisance indexes, and mapping of noise contours in space [51]. Noise indices are also used in contrast to the absolute measurements of sound pressure; these measurements are frequency weighted and time weighted with respect to exposure and allow assessment of perceived disturbance in a realistic manner. The deployment of UAVs in urban areas will have to incorporate noise indicators with geospatial population layers and habitat sensitivity to assist with new regulation enforcement and path planning. Life cycle assessment (LCA) offers a detailed model for assessing the cumulative environmental effect of the extraction of raw material through the manufacturing process, operation, and disposal of the product at the end-of-life. LCA measures the energy requirements, greenhouse gas emissions, resource consumption, and toxicity within the boundaries of the system [52]. In the case of UAVs, LCA implies that an important part of an embodied environmental footprint involves battery production as well as the production of composite materials and electronic components, which showcases why material circularity and material recycling are important. Eco-efficiency ratio is a ratio of function delivery (weight-haired-delivered, area covered, or service value) and environmental service (energy consumed, emission, or resource likeness). The multi-criteria optimization and trade-off analysis between the operational productivity and sustainability goals can be assisted by this ratio [53]. Mission sustainability indicators also incorporate reliability, safety margin, resource utilization efficiency, and environmental risk exposure into composite indices that are useful in the measurement of fleet-level performance benchmarking. Continuous computation of eco-performance metrics throughout the live operations is gradually facilitated by digital twins, telemetry analytics, and cloud-based dashboards.

Table 10.3 Core eco-performance metrics for UAV sustainability assessment

Metric	*Definition*	*Primary application*	*Limitation*
Specific energy consumption	Energy per payload-distance	Route optimization	Ignores life cycle
Carbon intensity	CO_2-equivalent per mission	Emission benchmarking	Data uncertainty
Noise index	Acoustic exposure	Urban compliance	Site dependency
Life cycle assessment	Full life cycle impact	Policy evaluation	High complexity
Eco-efficiency ratio	Output/environmental input	Multi-objective optimization	Requires normalization
Mission sustainability index	Composite indicator	Fleet benchmarking	Subjective weighting

With real-time monitoring, the adaptive control strategies can reduce energy peaks, eliminate excessive emission charging periods, and divert flights via low-noise corridors [54]. Nonetheless, interoperability is still a problem as a result of the variation of data format, sensor calibration, and the lack of standard reporting protocols (Table 10.3).

10.5 INTELLIGENT OPTIMIZATION FOR GREEN UAV OPERATIONS

In the quest to realize energy-efficient and environmentally friendly UAV operations, digital intelligence is progressively becoming the core element of the process. The emerging accessibility of onboard sensing, high-frequency telemetry, edge computing, and cloud connectivity allows the optimization of flight behavior and energy consumption, plus mission planning, to be continuous. Predictive decision-making and autonomous optimization of sustainability throughout the UAV life cycle are supported by artificial intelligence (AI), machine learning (ML), digital twins, and adaptive control systems [18]. Figure 10.4 depicts a closed-loop optimization ecosystem that involves the interactive combination of AI, digital twins, adaptive control, telemetry analytics, and power management.

10.5.1 AI-driven flight optimization and digital twin frameworks

Intelligent control technologies that utilize AI allow UAVs to optimize their trajectories, speed profiles, and altitude to reduce energy use and emissions without violating the restrictions of the mission. Optimum control policies are reinforcers that learn by evaluating historical flight data and simulated environments and adapt to erratic wind shear, variable payloads,

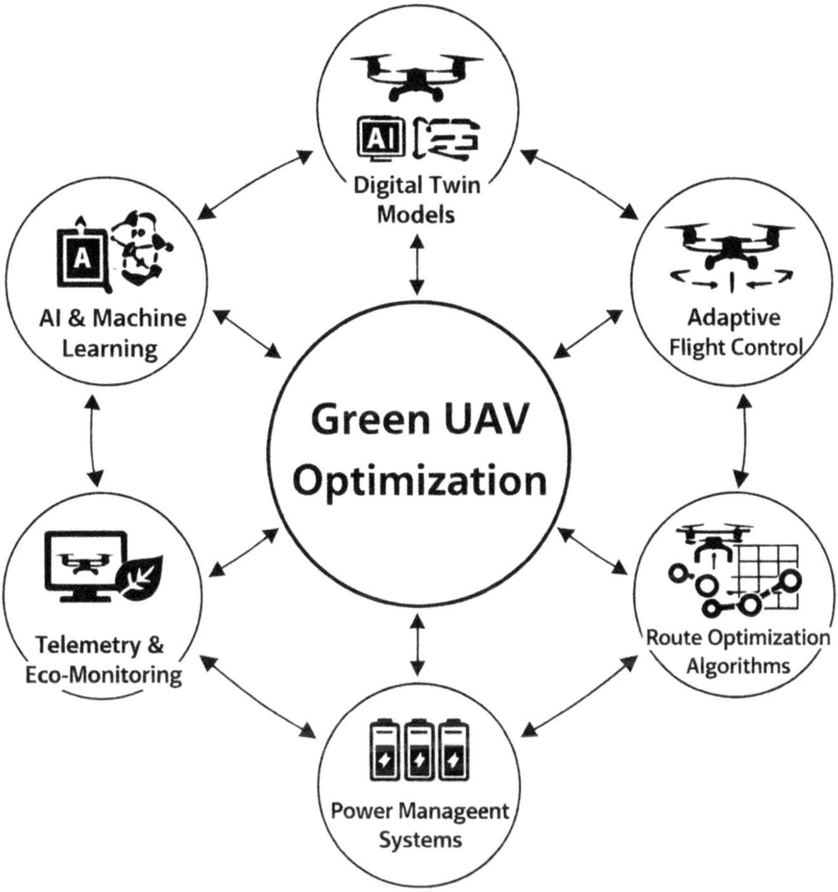

Figure 10.4 Intelligent optimization ecosystem for green UAV operations.

and battery state-of-health to create optimal control policies [55]. Predictive models are supervised based on the models of energy consumption and useful remaining flight time, which enhance reliability in missions and charging infrastructure scheduling [56]. Digital twins offer computerized approximations of operational UAVs through constructing the aerodynamic models, propulsion talent maps, battery degradation design, and environmental data channels [57]. The models allow testing scenarios, predicting their failure, and optimizing what-if without any physical asset being put at risk. By linking the digital twin with real-time telemetry, the constant calibration and refinement of the model are possible, and closed-loop optimization of the sustainability may be completed. The digital twins produce predictive analytics that help in fleet-level energy forecasting, preventive maintenance, and eco-performance benchmarking. Multi-objective optimization is a feature of route planning algorithms with a compromise between energy consumption, estimation of the flight time, avoidance of obstacles, regulatory restrictions, and landing areas into account [58]. Under uncertainty, graph-based search strategies, evolutionary algorithms, and swarm intelligence strategies are used to optimize complex airspace navigation in the city. Weather-aware routing has an additional benefit of minimizing energy wastage by taking advantage of good wind corridors and eliminating the inefficiencies caused by turbulence.

10.5.2 Adaptive power management and telemetry-driven eco-monitoring

A power management system is used to control the flow of power among batteries, motors, sensors, and communication modules in order to keep peak load and thermal stress at minimum levels [59]. Model predictive control systems are used to be able to allocate power in real time depending on the mission phase, the demand forecast, and battery conditions in order to maximize the life and extend the safety measures of components. Application of energy harvesting, such as solar augmentation, is an additional improvement to operational efficiency in the event of long-duration missions. Smart emergence eco-monitoring, based on telemetry, makes it possible to continuously measure power use, temperature, vibration, noise emission proxies, and degradation of the system [60]. The use of edge analytics also minimizes the latency and bandwidth needs, and cloud-based applications can provide sustainability reporting of the fleet and audit regulatory compliance. Between irregularities, faults of the components, and anomaly patterns in real time, anomaly detection algorithms are used to detect inefficiencies, component faults, and abnormal emission patterns. Schedulers made with AI are able to plan charging patterns according to the carbon intensity of the grid, the cost of electricity, and the availability of the fleet and minimize indirect emissions and the cost of operation [50]. Demand coordination is authorized by means of integration with the smart grid infrastructure and the use of renewable energy (Table 10.4).

Table 10.4 Intelligent optimization technologies for sustainable UAV operations

Technology	*Function*	*Sustainability benefit*	*Limitation*
Reinforcement learning	Trajectory optimization	Energy reduction	Training cost
Digital twins	Predictive modeling	Life cycle optimization	Model fidelity
Adaptive power control	Energy distribution	Battery longevity	Complexity
Telemetry analytics	Real-time monitoring	Early fault detection	Data volume
Smart charging	Grid-aware scheduling	Carbon reduction	Infrastructure

10.6 POLICY, STANDARDS, MATERIALS, AND FUTURE DIRECTIONS

A sustainable implementation of the UAV ecosystem will be identified based not only on technological innovation but also on controlled regulatory policies, unified environmental performance standards, and ethical management of materials. Such aviation regulatory bodies have been incorporating the environment into the certification of unmanned aircraft, which are run in more tolerant frameworks to comply with noise standards, operational safety, and carbon responsibility in the growing urban air transport tunnels [61]. International standardization is still a key issue to facilitate cross-border use of UAVs at the same time as maintaining similar sustainability levels and social acceptance [62]. Green standards of aviation focus on energy efficiency benchmarking, acoustic footprint handling, electromagnetic compatibility, and life cycle emissions disclosure. Transparent comparison is facilitated through voluntary sustainability reporting frameworks and environmental product declarations in order to make procurement decisions based on climate goals across platforms and suppliers of the UAV [63]. Introducing sustainability measures in airspace management systems also allows the use of eco-prioritized routing and the reduction of congestion. The challenge of material sustainability is becoming increasingly worrisome because the manufacturing of a UAV is dependent to a great extent on carbon fiber composites, lithium-ion batteries, and electronics with critical minerals. The recycling of composites is still too complicated and cost-restricted, and the end-of-life of the batteries needs to undergo scalability of lithium, cobalt, and nickel recoveries to address the depletion of resources and the leaking of the environment [10]. Bio-based composites, design-based structures of modular airframes, and principles of design-to-disassembly can enhance circularity and decrease the intensity of embodied carbon. Life cycle sustainability is also enhanced by battery second-life applications as well as closed-loop recycling infrastructures. The future UAV ecosystems will also be climate-sensitive mobility plans, which will include charging vehicles with renewable energy, the smart fleet solution, and noise-free operating paths. Traceability of material provenance using blockchain as a tool of certification of the material and automated compliance, aiding in transparent governance and responsibility [64]. New propulsion materials, such as solid-state batteries and hydrogen materials, improved by AI-based operational optimization, will have progressive effects on air mobility decarbonization, increasing the duration of missions in restricted environmental conditions. Long-term sustainability would involve the mutual investment between policy, industry, and research communities in the priorities to create interoperating standards, scalable networks of recycling, and an architecture of governance with data. The integration of these ecosystems will allow UAV platforms to have social importance beneficial to their environmental soundness and regulatory confidence (Figure 10.5 and Table 10.5).

Figure 10.5 Governance framework for sustainable UAV ecosystems.

Table 10.5 Sustainability governance dimensions for UAV ecosystems

Dimension	*Objective*	*Implementation focus*
Regulation	Safety and environmental compliance	Noise, emissions
Standards	Benchmarking	Eco-metrics
Materials	Circularity	Recycling
Infrastructure	Renewable integration	Smart charging
Governance	Transparency	Digital traceability

10.7 CONCLUSION

This chapter has investigated the dimensions that are crucial in energy conservation, emissions reduction, and eco-performance evaluation as the cornerstone to sustainable UAV ecosystems. The chapter provides a quantitative foundation for the assessment of the UAV sustainability by examining the physical determinants of energy consumption, comparative emission properties of new propulsion technologies, and embedded environmental performance metrics. The combination of artificial intelligence, digital twins, adaptive control, and monitoring supported by telemetry allows optimization of the behavior of the vehicles on the plane, the utilization of resources, and environmental competition on a continuous basis at both individual vehicle and fleet levels. Responsible deployment and long-lived resilience of the system are further enhanced using a policy framework, green aviation standards, and circular material strategies. All these interdisciplinary solutions contribute to the shift to the low-emission, energy-saving aerial mobility able to satisfy the regulations and sustainability requirements of society. With the increased use of UAVs in smart cities, logistics, and environmental services, it will be important to consider the practical components of eco-performance to be integrated during design, operation, and governance to ensure climate-relevant innovation and ensure that people trust autonomous aerial systems.

REFERENCES

1. Dorling, K., Heinrichs, J., Messier, G.G. and Magierowski, S., 2017. Vehicle routing problems for drone delivery. *IEEE Transactions on Systems, Man, and Cybernetics: Systems*, 47(1), pp. 70–85.
2. Gupta, L., Jain, R. and Vaszkun, G., 2021. Survey of important issues in UAV communication networks. *IEEE Communications Surveys & Tutorials*, 23(2), pp. 1123–1152.
3. Hassanalian, M. and Abdelkefi, A., 2017. Classifications, applications, and design challenges of drones. *Progress in Aerospace Sciences*, 91, pp. 99–131.
4. Otto, A., Agatz, N., Campbell, J., Golden, B. and Pesch, E., 2018. Optimization approaches for drone delivery. *Transportation Science*, 52(4), pp. 965–981.
5. Zeng, Y., Zhang, R. and Lim, T.J., 2019. Wireless communications with UAVs. *IEEE Communications Magazine*, 54(5), pp. 36–42.
6. D'Andrea, R., 2014. A revolution in the warehouse. *IEEE Transactions on Automation Science and Engineering*, 11(3), pp. 638–639.
7. Stolaroff, J.K. et al., 2018. Energy use and life cycle greenhouse gas emissions of drones. *Nature Communications*, 9, p. 409.
8. Hawkins, T.R. et al., 2013. Comparative environmental life cycle assessment of vehicles. *Journal of Industrial Ecology*, 17(1), pp. 53–64.
9. Ellingsen, L.A.W., Singh, B. and Strømman, A.H., 2016. Life-cycle emissions of electric vehicles. *Environmental Research Letters*, 11(5), p. 054010.

10. Peters, J.F., Baumann, M., Zimmermann, B., Braun, J. and Weil, M., 2017. Raw materials for lithium-ion batteries. *Renewable and Sustainable Energy Reviews*, 67, pp. 491–506.
11. Schäffer, B., Heutschi, K. and Wunderli, J.M., 2020. Drone noise emissions. *Applied Acoustics*, 165, p. 107347.
12. Christian, A.W. and Cabell, R., 2017. Psychoacoustic properties of UAV noise. *Journal of the Acoustical Society of America*, 141(5), pp. 3901–3911.
13. Torija, A.J. and Li, Z., 2021. Psychoacoustic annoyance of drones. *Applied Acoustics*, 170, p. 107534.
14. Gao, Y., Li, Y., Zhang, X. and Wang, Z., 2018. Recycling lithium-ion batteries. *Energy Storage Materials*, 13, pp. 1–11.
15. Raugei, M., Hutchinson, A., Morrey, D. and Winfield, P., 2018. Recycling of composite materials. *Resources, Conservation and Recycling*, 133, pp. 238–246.
16. Stolaroff, J.K. et al., 2018. Environmental implications of drone delivery. *Nature Communications*, 9, p. 409.
17. Pankaj, S. and Kumar, A., 2020. Sustainability metrics for UAVs. *Sustainable Computing*, 28, p. 100430.
18. Zhang, J., Cao, Y. and Wang, X., 2020. Energy-efficient UAV trajectory optimization. *IEEE Transactions on Wireless Communications*, 19(2), pp. 1214–1229.
19. Alzenad, M., El-Keyi, A. and Yanikomeroglu, H., 2018. 3D placement of UAV base stations. *IEEE Wireless Communications Letters*, 7(2), pp. 248–251.
20. European Union Aviation Safety Agency (EASA), 2022. *Environmental Protection in Aviation Report*. EASA.
21. Federal Aviation Administration (FAA), 2023. *UAS Integration Roadmap*. FAA.
22. Dorling, K. et al., 2017. Drone delivery routing. *IEEE Transactions SMC Systems*, 47(1), pp. 70–85.
23. Stolaroff, J.K. et al., 2018. Energy use and emissions of drones. *Nature Communications*, 9, p. 409.
24. Zeng, Y., Zhang, R. and Lim, T.J., 2019. UAV communications and energy tradeoffs. *IEEE Communications Magazine*, 54(5), pp. 36–42.
25. Austin, R., 2010. *Unmanned Aircraft Systems: UAV Design, Development and Deployment*. Wiley.
26. Verling, S. et al., 2020. Hybrid VTOL UAV energy tradeoffs. *Aerospace Science and Technology*, 98, p. 105689.
27. Anderson, J.D., 2016. *Fundamentals of Aerodynamics*. McGraw-Hill.
28. Gundlach, J., 2012. *Designing Unmanned Aircraft Systems*. AIAA.
29. Sun, Y. et al., 2019. Rotor interference effects in multirotors. *AIAA Journal*, 57(8), pp. 3412–3425.
30. Dorling, K. et al., 2017. Payload impacts on UAV routing. *IEEE Transactions SMC Systems*, 47(1), pp. 70–85.
31. Mueller, M.W. and D'Andrea, R., 2013. Quadrocopter dynamics. *IEEE Transactions Robotics*, 29(1), pp. 34–43.
32. Zhang, J. et al., 2020. Trajectory optimization for energy efficiency. *IEEE Transactions Wireless Communications*, 19(2), pp. 1214–1229.
33. Berecibar, M. et al., 2016. Battery degradation modeling. *Renewable and Sustainable Energy Reviews*, 56, pp. 572–587.

34. Traub, L.W., 2011. Range and endurance optimization. *Journal of Aircraft*, 48(3), pp. 703–711.
35. Peters, J.F. et al., 2017. Battery material sustainability. *Renewable and Sustainable Energy Reviews*, 67, pp. 491–506.
36. Plett, G.L., 2015. *Battery Management Systems*. Artech House.
37. Manthiram, A., 2017. Lithium–sulfur batteries. *Nature Communications*, 8, p. 15154.
38. Franco, C.D. et al., 2018. Wind impact on UAV energy. *Energy*, 159, pp. 834–845.
39. Jaguemont, J. et al., 2016. Temperature effects on Li-ion batteries. *Applied Energy*, 164, pp. 99–114.
40. Alzenad, M. et al., 2018. UAV placement and energy modeling. *IEEE Wireless Communications Letters*, 7(2), pp. 248–251.
41. Hawkins, T.R. et al., 2013. Life cycle assessment of electric vehicles. *Journal of Industrial Ecology*, 17(1), pp. 53–64.
42. Bradley, T.H. et al., 2009. Fuel cell aircraft systems. *Journal of Power Sources*, 193(2), pp. 794–803.
43. Noth, A., 2008. Solar-powered UAV endurance. *Autonomous Robots*, 24, pp. 237–249.
44. Schäffer, B. et al., 2020. Drone noise emission characteristics. *Applied Acoustics*, 165, p. 107347.
45. Zhang, J. et al., 2020. Energy-efficient UAV optimization. *IEEE Transactions Wireless Communications*, 19(2), pp. 1214–1229.
46. Torija, A.J. and Li, Z., 2021. UAV noise annoyance. *Applied Acoustics*, 170, p. 107534.
47. Otto, A. et al., 2018. Optimization for drone logistics. *Transportation Science*, 52(4), pp. 965–981.
48. Christian, A.W. and Cabell, R., 2017. Psychoacoustic UAV noise. *JASA*, 141(5), pp. 3901–3911.
49. Stolaroff, J.K. et al., 2018. Energy and emissions of drone delivery. *Nature Communications*, 9, p. 409.
50. Pauliuk, S. et al., 2019. Dynamic carbon accounting. *Journal of Industrial Ecology*, 23(3), pp. 646–658.
51. Schäffer, B. et al., 2020. UAV noise characteristics. *Applied Acoustics*, 165, p. 107347.
52. Hawkins, T.R. et al., 2013. LCA of electric vehicles. *Journal of Industrial Ecology*, 17(1), pp. 53–64.
53. Wursthorn, S. et al., 2011. Eco-efficiency indicators. *Journal of Cleaner Production*, 19 (15), pp. 1687–1698.
54. Zhang, Y. et al., 2021. Digital twins for UAV optimization. *IEEE Internet of Things Journal*, 8(7), pp. 5315–5327.
55. Kiumarsi, B. et al., 2018. Reinforcement learning in control systems. *IEEE Trans. Neural Networks and Learning Systems*, 29(9), pp. 4039–4050.
56. Liu, X. et al., 2021. Data-driven energy prediction for UAVs. *Energy*, 214, p. 118858.
57. Tao, F. et al., 2019. Digital twin-driven smart manufacturing. *IEEE Access*, 7, pp. 155653–155670.
58. Dorling, K. et al., 2017. UAV routing optimization. *IEEE Transactions SMC Systems*, 47(1), pp. 70–85.

59. Saini, H.K. and Jain, K., 2023, March 19. A novel deep energy efficient Hello packet scheduling for ad hoc networks in unmanned aerial vehicles. https://www.espublisher.com/journals/articledetails/843/
60. Zhang, Y. et al., 2021. IoT telemetry analytics. *IEEE Internet of Things Journal*, 8(7), pp. 5315–5327.
61. Saini, H.K. and Jain, K., 2024. Implementation of DEWMA-based hello packet for AODV to improve the performance of FANET with 3D-GMM. *SN Computer Science*, 5, p. 321. https://doi.org/10.1007/s42979-024-02673-z
62. International Civil Aviation Organization (ICAO), 2021. *Environmental Trends in Aviation*. ICAO.
63. Saini, H., 2025. GeNp-ODHR: Green energic and smart network performance estimation-based optimized deep hello routing for flying Ad Hoc network. *SAE International Journal of Connected and Automated Vehicles*, 8(3), pp. 319–335. https://doi.org/10.4271/12-08-03-0022
64. Kshetri, N., 2018. Blockchain's roles in strengthening cybersecurity and protecting privacy. *Telecommunications Policy*, 41(10), pp. 1027–1038.

Chapter 11

Automated agricultural spraying using commercial UAV platforms

Sarumi Usman Abidemi, Matin Taghian Zadeh, Cliford Peter Mlimwa, and Fadi Al-Turjman

11.1 INTRODUCTION

11.1.1 Motivation and importance

There is an increasing pressure in the whole global agricultural industry to increase food production with fewer available resources (Garcia-Munguia et al. 2024). The current situation of agricultural operations shows that there is an acute necessity for innovative solutions that can prove their intelligence and efficiency and be scaled to the necessary size. Precision agriculture is a groundbreaking kind of agricultural system, which combines the use of modern technology and conventional farming techniques to enhance crop yield and reduce wastage and uphold environmental improvement. At present, DJI is one of the largest vendors of automated spraying drones that promise a lot of potential as a technological solution (Hafeez et al. 2023). The drones deliver precision spraying through their ability to generate field maps and perform terrain and crop evaluations before executing automated applications that eliminate excess chemicals and minimize labor requirements and operational spending, as shown in Figure 11.1.

Drone technology giant DJI (DJI Agriculture, 2023) leads the market through its development of agricultural drones known as the DJI Agars series, which operate through high-resolution sensors, AI-based software, and GPS-based navigation and automated flight planning for smart spraying operations, as indicated in Figure 11.2.

11.1.2 Scope of the survey

- This chapter reviews how automated spraying drones operate in precision agriculture through a specific examination of DJI agricultural drones as a dominant commercial solution. The review analyzes the following domains:
- The drone-based land mapping and terrain analysis technologies for spraying operations.

DOI: 10.1201/9781003731580-11

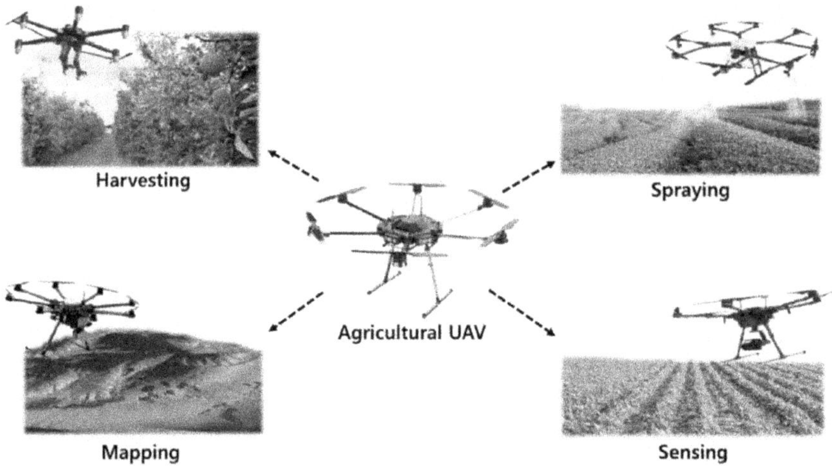

Figure 11.1 Multifunctional applications of agricultural UAVs in precision farming.

Figure 11.2 DJI Agras T40 drone spraying pesticide over a crop field.

- The sensors, along with AI algorithms that enable drones to adjust their spraying patterns according to field conditions.
- The research evaluates both the effectiveness of spraying techniques (Borikar et al., 2022)and their environmental consequences.

- The study sorts existing research according to the technology used, application domain, and methodology approach.
- The chapter identifies current obstacles facing drone-based agriculture, which include battery capacity issues alongside payload weight restrictions, economic costs, and processing speed.
- The chapter predicts the upcoming trend of connecting agricultural drones with AI and IoT technologies through cloud-based systems.

This literature review examines material from 2016 to 2025 to present the most current developments in field implementations. The main analysis of DJI drones includes side-by-side assessments with alternative drone technologies and systems.

11.1.3 Survey objectives

- Survey Objectives: The survey chapter pursues five primary objectives.
- The chapter delivers an extensive examination of drone-based agricultural spraying technologies through its analysis of the sector's complete range.

11.1.4 Organization of the chapter

This survey chapter is organized as shown in Figure 11.3.

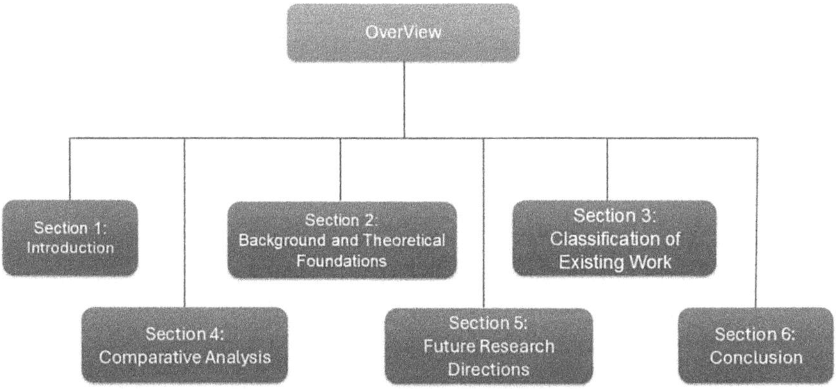

Figure 11.3 Diagram of the chapter overview.

11.2 BACKGROUND AND THEORETICAL FOUNDATIONS

11.2.1 Introduction to agricultural UAVs

Unmanned Aerial Vehicles (UAVs), commonly called drones, are now being used much more often in agriculture. Because they are now used in modern farming, farmers have access to efficiency, lower costs, and real-time choices. In farming, drones help with aerial surveys, soil assessment, regular crop observation, spraying pesticides, and figuring out what the yield will be (Önler et al. 2023).

It was in the 1980s in Japan that agricultural UAVs were first used for experimental spraying in rice fields. Thereafter, improvements in sensors, GPS, computer vision, AI, and batteries for drones have made smart farming rely on them.

11.2.2 Precision agriculture is built on several theoretical backgrounds

Precision agriculture is when farming is organized using data based on changing conditions over different parts of a field. The purpose is to reach the highest possible return without wasting resources. This idea depends on UAVs, which send real-time data into smart systems that help with decisions on the battlefield (FAO 2021).

Prominent theories that explain anxiety are the following:

- Variable Rate Technology (VRT) uses UAV information to apply fertilizers or pesticides only where needed.
- Remote sensing theory consists of understanding reflections from cameras (mostly visible, IR, and multispectral) to determine how healthy the crops are.
- Sensor data is used by these systems to guide or make automatic decisions (for instance, adjust the sprayer rate).

Because of these frameworks, UAVs can serve in surveillance as well as in closed-loop control systems used in farming.

11.2.3 Key components of agricultural UAV systems

Some of the main elements of agricultural unmanned aerial vehicles include their motors, electronics, and sensors (Saini and Jain 2023, March 19).

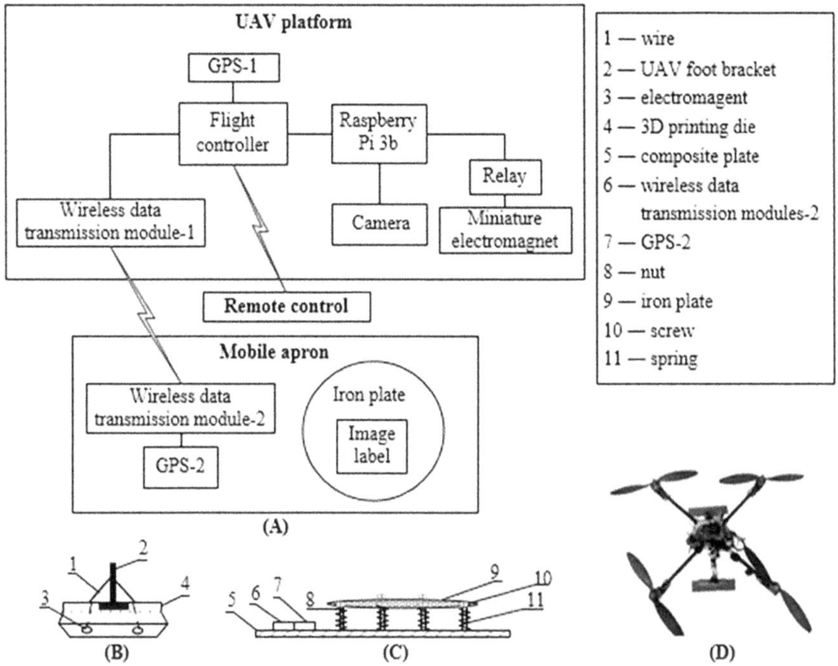

Figure 11.4 UAV subsystems diagram.

Figure 11.4 illustrates that the integrated subsystems are included in agricultural UAVs, such as

- The flight control system helps to keep the drone level and directs it with the help of GPS, IMU, and a barometer.
- Modern drones contain cameras (RGB, multispectral, thermal), sprayers, and LiDAR sensors as key parts of their payload system.
- Using a communication system, the operator can take control instantly, get telemetry back, and watch video feeds.
- With mission planning software, field maps are used by users to map out paths and organize tasks automatically.

11.2.4 Sensor technologies in UAVs for agriculture

Using sensors on UAVs for agriculture. How well a UAV flies depends greatly on the sensors it has. Common sensors for detecting chemical reactions are

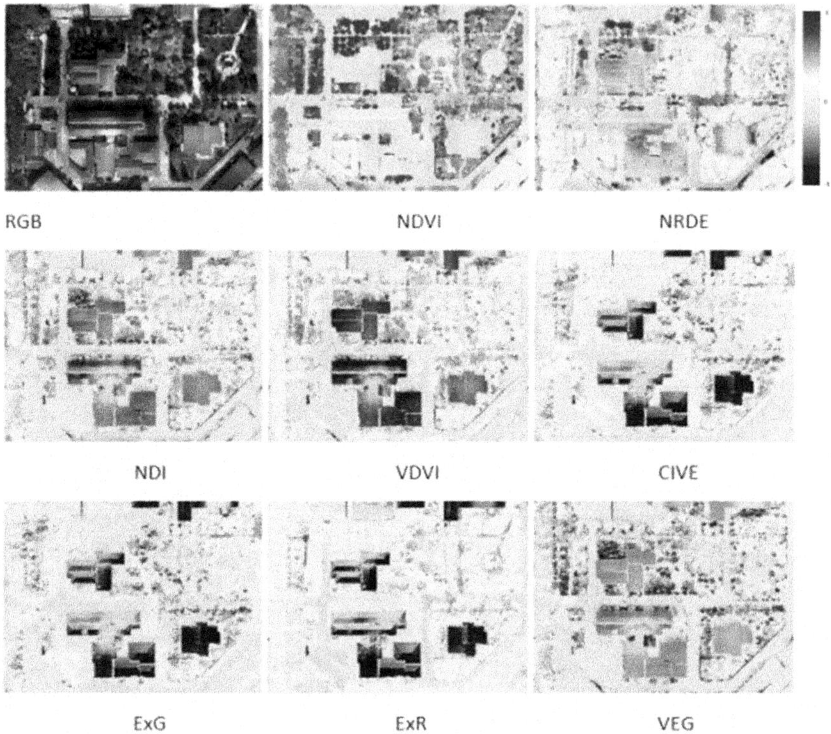

Figure 11.5 Images from different UAV sensors.

RGB cameras, which are used for routine imaging and checking products with the eye. Plant stress can be identified using multispectral sensors and indices like normalized difference vegetation index (NDVI). Using thermal cameras allows me to spot areas of water stress and any diseases. LiDAR makes it possible to get precise 3D maps and canopy models. With the data they gather, these sensors check crop health, notice unusual events, and predict future outcomes (Hafeez et al. 2023) (Figure 11.5).

11.2.5 Data processing and analysis techniques

Processed data is necessary for UAV data collected during operations. Most of the time, this is possible thanks to machine learning (ML) or deep learning (DL) models (Hasan et al. 2020).

- Drones produce images that are blended through photogrammetry, which produces a detailed map.
- Segmentation algorithms distinguish where the crop is from where the background is to determine how much area it covers.

- Object detection models help find weeds, pests, or abnormalities on plants.
- Changes in crop growth can be tracked over time to estimate how much each crop yield will be.

11.2.6 How AI is used in drones for agriculture

AI is very important in helping UAVs become more autonomous and smart. Three common ways are the Dijkstra and RL-based algorithms for finding the best path. Drone Simultaneous Localization and Mapping (SLAM) and computer vision technology enable drones to see and dodge objects in front of them. AI can identify if, when, and how much pesticide or fertilizer is needed depending on what the crop or soil looks like (Huang et al. 2009).

11.3 CLASSIFICATION OF EXISTING WORK

11.3.1 Overview

Many studies have been done on using UAVs in agriculture by combining different sensors, AI systems, and ways to process data. To make things easier to understand, we divide existing work into major categories called dimensions based on area as shown on Table 11.1. Table 11.2 makes classification based on season type, and Table 11.3 based on AI/ML technique.

1. Application area,
2. Sensor type,

Table 11.1 Classification based on application area

Application	*Description*	*Representative works*
Crop monitoring	Using UAVs to observe plant health, growth stage, or stress.	García-Munguía (2024), Hafeez et al. (2023)
Weed detection	Identifying and localizing weeds for site-specific treatment.	García-Munguía (2024), FAO (2021)
Pest and disease detection	Spotting signs of infestations or infections early.	Saini and Jain (2023, March 19)
Soil analysis	Estimating moisture, pH, or nutrient levels from aerial data.	Hafeez (2023)
Yield prediction	Forecasting productivity using time-series data and AI.	Hassan et al. (2020)
Spraying and seeding	Using drones to deliver inputs precisely.	Haung et al. (2009), Li et al. (2020)

Table 11.2 Classification based on season type

Sensor type	*Purpose*	*Example*	
RGB camera	Basic visual inspection and mapping.	Detecting growth irregularities.	Önler et al. (2023)
Multispectral sensor	Plant stress detection, NDVI calculation.	Monitoring chlorophyll content.	Lin et al. (2022)
Thermal camera	Water stress detection, canopy temperature.	Spotting drought-affected zones.	Mogili and Deepak (2018)
LiDAR	3D terrain and canopy modeling.	Topographical mapping of fields.	Özyurt et al. (2022)
Hyperspectral camera	Advanced stress and nutrient detection.	Early disease diagnosis.	Li et al. (2020)

Table 11.3 Classification based on AI/ML technique

Technique	*Model used*	*Use case*	
Traditional ML	SVM, Random Forest, KNN	Özyurt et al. (2022)	Rejeb (2022)
Deep learning	CNN, ResNet, YOLO, UNet	Object detection, segmentation	Zhang and Kovacs (2012)
Time-series models	LSTM, GRU	Growth prediction	ÖNLER (2023).
Reinforcement learning	DQN, PPO	Navigation and path optimization	Özyurt et al. (2022)
Hybrid models	CNN + LSTM, Sensor Fusion + ML	Combining spatial and temporal learning	Sankaran (2021)

3. Machine learning approach,
4. The amount of decision-making power given to the workers,
5. The ability to see things in space and time.

Using this system helps researchers locate the weaknesses, highlights the strong points, and outlines further prospects in UAV-based agriculture as shown in Table 11.1.

High precision jobs in detecting pests or parts of plants often use YOLO and UNet which are built on deep learning (Table 11.4).

Table 11.4 Classification based on autonomy

Level autonomy	*Description*	*Example*	
Manual	Human-controlled flight and decisions.	Early-stage UAV systems	García-Munguía (2024)
Semi-autonomous	Autonomous data capture, manual analysis.	Most current deployments	Saini and Jain (2023, March 19)
Fully autonomous	AI handles navigation, data processing, decision-making, and action.	Advanced research systems	Hafeez (2023)

11.3.2 Classification based on data resolution

The sensor can only detect objects that are a certain size or larger, which is called the spatial resolution. Unmanned aerial vehicles have a better spatial resolution (a few centimeters for each pixel) than satellites.

Temporal Resolution Measures How Often Collected Data Is Obtained: UAVs give reliable daily or sometimes hourly images, which help to monitor crops closely.

Using datasets with higher resolution usually makes the model more accurate, but it can be difficult to process them and requires strong hardware.

11.4 COMPARATIVE ANALYSIS OF APPROACHES

11.4.1 Strengths and weaknesses of various approaches

Farmers using unmanned aerial vehicles in agriculture have tried out different technologies and ways of doing things. I am now comparing each statistic below:

RGB Imaging + ML. It is less expensive, widely used, and straightforward to deploy. It only covers a small range of light wavelengths, so stress detection is less precise (FAO 2021).

Any observation done with multispectral imaging and learning can locate areas of plant health concerns (NDVI), but the equipment is more expensive and requires adjustments (García-Munguía 2024). For example, YOLO has high accuracy and detects objects quickly, but you need a large dataset and a GPU.

In terms of semantic segmentation (using UNet), locating specific areas within the image is done through exact classification; however, the method is computationally expensive and sensitive to lighting.

Most traditional ML methods (SVM and RF) are suitable for handling small data, and you can interpret the results easily.

Sensor fusion has a high tolerance for mistakes and uses different types of input. This type of spraying only works with pre-programmed rules. This approach is simple to set up but does not handle unexpected events well. AI systems used in spraying (for example, DJI Agras) save resources, are versatile, and may be too costly for some farmers (Hassan et al. 2020).

11.4.2 Trends and patterns in research

Most recent studies have moved away from normal ML and toward CNNs or hybrid DL because they excel in image analysis and grouping.

More and more, multispectral and hyperspectral sensors are being used, mainly to identify diseases and detect stress. With smaller droplets, chemicals are applied more accurately.

Studies are now working on putting models directly on drones using small computers (e.g., Jetson Nano, Raspberry Pi), allowing drones to process data without using ground stations.

Precision spraying drones by DJI and similar firms now rely on automated, adaptive operations to control and direct the aircraft (Hafeez 2023).

Processing, analyzing, and saving UAV data are done more and more with the help of cloud technologies. This is because most of the farms can be closely monitored.

The lack of data is managed by publishing open datasets (including Agriculture-Vision) by means of simulation (e.g., AirSim) (Hafeez et al., 2023).

11.4.3 Open challenges and research gaps

Nevertheless, despite the improvements, there are still certain issues:

Annotated agricultural data is not abundant, primarily due to deep learning, and thus, it is difficult to do so.

generalize results in other circumstances. Models trained for one region may not work in other regions since the soil, growing plants, and weather there are not the same.

You might not understand how deep models make decisions, which can lead to wrong decisions in agriculture. Because of their power supply, UAVs are restricted in the kinds of tasks they can perform. It remains very difficult from a technical standpoint to ensure that data from various sensors is fused correctly and reliably. Restrictions on drone flights are found in various regions, mainly over private farmland, because of privacy concerns. Most models are trained in advance and not while data is being collected. Adapting to sudden changes in the field is not often seen but is very important for success.

11.5 FUTURE RESEARCH DIRECTIONS

11.5.1 Emerging trends

A number of promising advances are guiding where autonomous drone use in agriculture is heading.

- Technology Companies are Applying Edge AI and Onboard Intelligence: Many are interested in sending lightweight AI straight to drones. Through devices like the NVIDIA Jetson Nano and the Raspberry Pi 4, it is now possible to do real-time object detection, make maps of places, and automate decisions without the help of a central server (Saini and Jain 2023, March 19).
- Making Federated Learning Useful in Agricultural AI: Because of federated learning, several farms can take part in learning together without exposing their data (Zhang and Kovacs 2012).
- Advanced Technologies Used: By merging information from thermal, hyperspectral, LiDAR, and RGB technologies, farmers will get more precise information about crops, which will allow them to use the spray with precise coverage (Sankaran 2021).
- The ability to Work with IoT and Smart Farming Systems: Drones will be involved in more areas, cooperating with weather stations, soil sensors, and autonomous vehicles as part of a connected precision agriculture system (Saini and Jain 2023, March 19).
- Swarm Drones: Having many drones flying together (swarming) allows large-scale farms to cover more space, be more efficient, and adjust more quickly. They are already being worked on and tested in China and Japan (Rejeb 2022).

11.6 CONCLUSION

Using drones in agriculture, especially for spraying pesticides, is a major shift in how farming is done. The chapter has examined the current state of intelligent agricultural spraying drones, primarily devoted attention to the systems of DJI, and has investigated the broader effects of the application of artificial intelligence, data combination, and automation of crop management. The main technologies are cut by the background sections and approaches that are real-time kinematics (RTK), AI-based decision-making, and vegetation classification with computers. It resulted in a basis of understanding on the ways UAVs perceive the land, study conditions of crops, and perform spraying work independently in various circumstances.

It was found that this topic encompasses a wide range of topics, with a low and a high end, as shown by the review of the published works, types based on technology, including machine learning and rule-based systems similar to multispectral imaging and onboard edge computing, and lastly, application types such as herbicide spraying, water conservation, and pest control. This method made it possible to see clearly the extent of development of the field and the amount and breadth of its activities. Based on our research, UAVs tend to perform well in environments where their factors are known, yet have challenges with change of things. The primary problems found were limited battery life, changing weather conditions, no single standard, and a low capability of drones to make decisions. Other critical advantages, such as terrain conscious mapping, accurate spraying, and fewer, were also present, which implied that they could be applied on a higher level and in the market.

REFERENCES

Borikar, G. P., Gharat, C., & Deshmukh, S. R. (2022). Application of drone systems for spraying pesticides in advanced agriculture: A review. *IOP Conference Series: Materials Science and Engineering*, 1259(1), 012015.

DJI Agriculture. (2023). *Agras T40 drone product page*. DJI. https://www.dji.com/agras-t40

FAO. (2021). *The state of food and agriculture 2021 – Making agrifood systems more resilient to shocks and stresses*. Food and Agriculture Organization of the United Nations. https://www.fao.org/documents/card/en

García-Munguía, A., Guerra-Ávila, P. L., Islas-Ojeda, E., Flores-Sánchez, J. L., Vázquez-Martínez, O., García-Munguía, A. M., & García-Munguía, O. (2024). A review of drone technology and operation processes in agricultural crop spraying. *Drones*, 8(11), 674.

Hafeez, A., Husain, M. A., Singh, S. P., Chauhan, A., Khan, M. T., Kumar, N., ... Soni, S. K. (2023). Implementation of drone technology for farm monitoring & pesticide spraying: A review. *Information Processing in Agriculture*, 10(2), 192–203.

Hasan, K. M., Hasan, M. T., Newaz, S. S., & Ahsan, M. S. (2020). Design and development of an autonomous pesticides spraying agricultural drone. In *2020 IEEE Region 10 Symposium (TENSYMP)*, Dhaka, Bangladesh (pp. 811–814). IEEE.

Huang, Y., Hoffmann, W. C., & Lan, Y. (2009). Development of a spray system for an unmanned aerial vehicle platform. *Applied Engineering in Agriculture*, 25(6), 803–809.

Li, C., Liu, F., Zhou, X., & Wang, J. (2020). Deep learning for UAV-based plant disease recognition: A review. *Sensors*, 20(23), 6761. https://doi.org/10.3390/s20236761

Lin, Y., Ma, R., Liu, Z., & Li, M. (2022). Intelligent control and path planning for agricultural UAV spraying operations. *Computers and Electronics in Agriculture*, 199, 107106. https://doi.org/10.1016/j.compag.2022.107106

Mogili, U. R., & Deepak, B. B. V. L. (2018). Review on application of drone systems in precision agriculture. *Procedia Computer Science*, 133, 502–509. https://doi.org/10.1016/j.procs.2018.07.063

Önler, E., Özyurt, H. B., Şener, M., Arat, S., Eker, B., & Çelen, İ. H. (2023). Spray characterization of an unmanned aerial vehicle for agricultural spraying. *The Philippine Agricultural Scientist*, 106(1), 39–46.

Özyurt, H. B., Duran, H., & Çelen, İ. H. (2022). Determination of the application parameters of spraying drones for crop production in hazelnut orchards. *Tekirdağ Ziraat Fakültesi Dergisi*, 19(4), 819–828.

Rejeb, A., Simske, S., Keogh, J. G., & Rejeb, K. (2022). Smart farming through emerging technologies: A review of AI, IoT, and blockchain. *Technological Forecasting and Social Change*, 178, 121603. https://doi.org/10.1016/j.techfore.2022.121603

Saini, H. K., & Jain, K. (2023, March 19). A novel deep energy efficient Hello packet scheduling for ad hoc networks in unmanned aerial vehicles. https://www.espublisher.com/journals/articledetails/843/

Sankaran, S., Khot, L. R., & Espinoza, C. Z. (2021). UAS-based pesticide application in agriculture: Opportunities and challenges. *Agricultural Systems*, 191, 103149. https://doi.org/10.1016/j.agsy.2021.103149

Zhang, C., & Kovacs, J. M. (2012). The application of small unmanned aerial systems for precision agriculture: A review. *Precision Agriculture*, 13(6), 693–712. https://doi.org/10.1007/s11119-012-9274-5

Chapter 12

Drones application in precision agriculture

A survey of techniques, challenges, and opportunities

Sarumi Usman Abidemi, Zainab Babangida Umar, Cliford Peter Mlimwa, and Fadi Al-Turjman

12.1 INTRODUCTION

12.1.1 Motivation and importance

Precision agriculture is the use of technologies that combine information systems, intelligent devices, sensors, and structured management to improve the performance of crops and sustainability in agriculture (Veroustraet, 2015). Precision agriculture is important because it helps farmers grow more crops and improve their production. It also helps protect the environment. Since farmers use only what is needed, there is less waste and pollution. Precision agriculture also improves working conditions for farmers and can support better care for animals on the farm. In short, precision agriculture makes farming more efficient, profitable, and eco-friendly.

The success of precision agriculture relies on advanced technologies like GPS, GIS, IoT, AI, and remote sensing. GPS makes it possible to map agricultural fields precisely and gives agricultural equipment precise location data. GIS integrates soil maps, weather information, satellite images, and other data layers, making it possible for farmers to gain knowledge about field variability and make good decisions by spatially visualizing this data. IoT gathers real-time data from crop monitoring devices on the environmental conditions (Khose et al., 2023; Mekonnen et al., 2019; Celen et al., 2020).

Additionally, drones can be used to spray chemicals in a specific region, saving time, reducing waste, and protecting the environment (Caballero et al., 2019). Figure 12.1 illustrates a sample of drones that are a useful tool for modern precision agriculture since they save time, reduce expenses, and boost crop yields. This study reviewed the applications and importance of using drones/UAVs in agriculture, as well as the difficulties and future improvements. The chapter is organized as follows:

- Section 12.1 discusses the background information.
- Section 12.2 presents the literature review, which is divided into (a) an overview of drone types and technologies used in agriculture and (b) key applications of drones in precision farming.

 DOI: 10.1201/9781003731580-12

Figure 12.1 Drone in precision agriculture.

- Section 12.3 explains the research methods used in this research.
- Section 12.4 discusses the main findings.
- Section 12.5 concludes the chapter with final thoughts and possible future work.

12.2 LITERATURE REVIEW

12.2.1 Drone types and technology in agriculture

Drones used in precision agriculture are usually divided into two main categories: fixed-wing and multirotor UAVs. Each has its own unique advantage and limitations depending on the task it's used for.

Fixed-wing drones are made more like airplanes, with wings that create lift as they fly. They are far more effective for long distances since they don't need to hover. Many types can fly for more than an hour, and some that use petrol motors rather than batteries may fly for up to 16 h. They are therefore useful for swiftly mapping vast areas or covering huge farms. Also, some of them can support loads of up to 50 kg, which is good for some sensors or payloads (Gar, 2022). They mostly require unique settings and a lot of space to take off and land. They are not great in small spaces or for detailed inspection work because they can't hover or make sharp turns easily. Additionally, they are typically more costly and heavier than multirotors; however, these costs have lately decreased.

Quadcopters are the most used drones in both commercial and recreational farming among multirotors. They are easier to fly, more accessible, cheaper, and might stay in place. They are good for closer checks, aerial photography, and crop monitoring, particularly in small or rough areas. They are also easy to use because they only need a little space to take off or land (Thamm et al., 2015). Compared to fixed-wing drones, as shown in Figure 12.2, most can only fly for 20 to 30 min and only cover a small area. As such, they are not ideal for big farms and remote locations (Vohra et al., 2024). But, in terms of flying duration and range, quadcopters, as shown in Figure 12.3, have a significant disadvantage. They use electric motors, such

Figure 12.2 Fixed-wing drone.

Figure 12.3 Quadcopter drone.

as a single motor drone, as shown in Figure 12.4, and a hexacopter drone, as shown in Figure 12.5, which can be weather-sensitive and require regular service because of their complex structure. Overall, multirotors are a more suitable option for situations that require precision and ease of use, like in small farms, orchards, or places requiring hovering and close observation.

However, large-scale mapping, wide area surveys, and tasks requiring heavy equipment and longer flying times are better suited for fixed-wing drones (Vohra et al., 2024).

Figure 12.4 Single rotor drone.

Figure 12.5 Hexacopter drone.

12.2.2 Sensor types

Drones have different sensors like multispectral, thermal, RGB cameras, and LiDAR systems. These sensors help farmers manage plant health, soil, and the condition of the field. For example, multispectral cameras can detect things like chlorophyll and leaf water content and are used to calculate vegetation indices such as normalized difference vegetation index (NDVI) (Daponte et al. 2019; Berni et al., 2009; Burgos et al., 2015).

12.2.3 Applications of drones in precision agriculture

Santesteban et al. (2017) and Primicerio et al. (2012) focused on using a custom-designed UAV with a high-resolution thermal camera to monitor water stress in a commercial vineyard. These Crop Water Stress Index (CWSI) values were compared with on-ground measurements like stem water potential and stomatal conductance at 14 sample sites. The CWSI had a good correlation with these field measurements, making it a reliable tool to detect instantaneous water stress. Multiple flights would be required over time to adequately capture seasonal fluctuation, since, although the thermal data also offered valuable information at a seasonal level, it didn't always match long-term trends in water condition.

FREYR Drone, a quadcopter designed to spray pesticides over farmland, was presented by Spoorthi et al. (2017). Its objective is to reduce farmers' workload and save time by spraying evenly across the entire field. The drone is controlled on an Android application, which is connected to a Wi-Fi module, and it uses GPS to follow the shape of the field. The system includes an Arduino board, sensors such as an accelerometer, gyroscope, and magnetometer, which provide stability, and a wireless camera for high-quality images. The goal is to provide farmers with an automated spraying solution that is easy to use.

In Shaw and Vimalkumar (2020), the authors describe the development of an octocopter drone, which is designed for pesticide spraying. The idea is to lower the health risks and labor involved in manual spraying, which can cause serious health issues. The drone contains nozzles to produce a fine spray of water over crops, a 6-liter tank, and a pump. It has enough power to carry large loads and spray effectively due to its 8 BLDC motors. The drone can fly with a total force of 38.2 kg and comes with a solid landing gear. In addition to agriculture, the drone might be used to disinfect cities. It lowers labor expenses, saves time, and lowers the cost of applying pesticides.

The study by Chin et al. (2023) reviewed academic publications to investigate drone types, disease types, crops, machine learning tasks, and data sources. The commonly studied diseases and fungal infections are the main focus. Grapes were frequently analyzed plants. The common drone used was the quadcopter, and classification was the main task, often using CNN models. Most studies used field images, especially color infrared, for detecting disease symptoms. In Shaw and Vimalkumar (2020), early and accurate detection of tomato plant diseases using hyperspectral imaging from UAVs was studied. The study targeted two diseases with similar early symptoms: target spot and bacterial spot. They used two classification methods, a multilayer perceptron and stepwise discriminant analysis. Important wavebands were in the blue, red, and red-edge ranges. The MLP method gave a very high classification accuracy of 99%. This shows that UAVs with hyperspectral sensors can be a powerful tool for plant disease diagnosis in real conditions.

An AI-based spraying drone was developed to detect and spray cherry trees more efficiently by Akdogan et al. (2024). Hassan et al. (2022) study focuses on detecting and counting rice plants using drone-based aerial imagery and GIS to improve crop yield.

Kitano et al. (2019) present an approach that combines computer vision techniques with transfer learning using CNNs. It was tested on spinach plants in an experimental field. The method achieved a 95% accuracy for counting plants in small areas with high image resolution. Even when the resolution was reduced by half, the extra error was just 0.7%. However, when applied to the full field, the error increased to 42.5%, still showing the method's practical potential.

A framework for monitoring and yield estimation in pomegranate plantations using a quadcopter was proposed in Duggal et al. (2016). The drone automatically flies through rows of crops by following GPS waypoints planned through convex optimization. The navigation system was built on the robot operating system and generates fast and free collision paths. A new dataset of pomegranate growth stages with videos and labeled images was also developed as part of the work. The drone uses supervised learning to estimate yield from the collected images. The system is currently focused on pomegranates but can be adapted to other crops. Future work includes adding near-infrared sensors for better predictions and even monitoring livestock within the fields.

In Abdulridha et al.'s (2020) paper, drones and lab setups using hyperspectral imaging were used to detect tomato plant diseases. They used two types of tomato plants, one at risk for TYLC and the other able to tolerate, and conducted tests both in the lab and in the field with a UAV. The study by Albetis et al. (2017) looks at the use of UAV multispectral imaging to detect grapevine disease. Drones fitted with visible and near-infrared cameras were used in this investigation. Ground truth data were gathered to compare healthy and diseased plants. Both univariate and multivariate classification techniques were examined using 20 image-based variables (biophysical parameters, vegetation indices, and spectrum bands). Findings show that red cultivars had a high accuracy rate for symptom identification, but white cultivars had a lower detection rate. The researchers came to the conclusion that UAV photography still has potential for mapping disease symptoms in vineyards, notwithstanding a few misclassifications.

A smart agriculture drone system for crop monitoring and spraying that integrates IoT and machine learning is presented in the Berner and Chojnacki (2017) study. The drone recognizes three different types of crops from a trained dataset using TensorFlow Lite and the EfficientDet-Lite1 model. It can shift between full- and half-capacity spray modes based on what it detects. By using preset locations and applying pesticides only when necessary, it works independently to lower chemical use and promote sustainable farming. Meivel and Maheswari (2021) work focuses on using drone-based sensors to help plant growth monitoring and affordable, organic spraying systems. The authors compare two methods: a drone system that uses GPS and a neural network-based approach that uses Quantum GIS (QGIS). Using a drone fitted with NIR and NDVI sensors, multispectral photos of farmland were taken by Gnädinger and Schmidhalter (2017).

12.3 RESEARCH METHODOLOGY

The research was based on reviewing existing studies about how drones are used in precision agriculture. Multiple papers that explain how drones help in modern farming were collected and analyzed, especially in tasks like managing crops, detecting diseases, spraying, and data collection. This method is selected because it makes it possible to look into different kinds of information that have already been studied by experts in the field.

The information was gathered by searching for research papers on academic websites and databases, which include IEEE Xplore, ScienceDirect, Springer Nature, and Google Scholar. The search keywords are: drones in agriculture, UAV crop monitoring, smart farming with drones, and drones in precision agriculture.

The chosen papers were published between 2015 and 2025. To make sure about the reliability of the information, only papers that were peer-reviewed, meaning they were checked by other researchers before being published, were included. Sources that are not academic, like blogs or articles, were ignored. Studies about drones used for military or non-agricultural purposes are not included. The papers were organized into three categories: (1) drone types, (2) types of sensors used, and (3) application of drones in farming.

A lightweight UAV-based remote sensing system was created by Jeong et al. (2016) to track the growth of rice crops under different nitrogen treatments. Nitrogen-treated rice paddies were used in the experiment. The fields were photographed using a camera. They used atmospheric correction to increase accuracy. According to the study's findings, UAV-based multispectral photography is an easy and effective way to obtain trustworthy crop growth data and may be used to track changes in field conditions. Several types of drones are shown in the comparison Table 12.1.

Table 12.1 Different types of drones and their applications

Author(s)	*UAV type*	*Application*
Primicerio et al. (2012)	Hexacopter UAV	Crop Monitoring and Health Assessment
Santesteban et al. (2017)	Multirotor	Soil and Irrigation Analysis
Spoorthi et al. (2017)	Quadcopter	Pesticide Spraying and Seeding
Shaw and Vimalkumar (2020)	Octocopter	Pesticide Spraying and Seeding
Chin et al. (2023)	Quadcopter	Pest and Disease Detection
Shaw and Vimalkumar (2020)	UAV with hyperspectral imaging	Pest and Disease Detection
Akdoğan et al. (2024)	AI-powered quadcopter	Pesticide Spraying and Seeding
Hassan et al. (2022)	UAV with GIS and image processing	Crop Counting and Soil Health Estimation
Hassan et al. (2022)	Quadcopter	Plant Counting and Yield Estimation
Kitano et al. (2019)	Quadcopter	Plant Counting and Yield Estimation, Crop Monitoring and Health Assessment
Duggal et al. (2016)	Hexacopter	Pest and Disease Detection
Albetis et al. (2017)	Quadcopter	Pest and Disease Detection
Berner and Chojnacki (2017)	Quadcopter (X500 frame)	Pesticide Spraying and Seeding, Crop Monitoring and Health Assessment
Meivel and Maheswari (2021)	Quadcopter	Irrigation Management, Organic Spraying, Vegetation Health Monitoring
Gnädinger and Schmidhalter (2017)	Octocopter UAV	Precision Phenotyping, Plant Counting, Nitrogen-

12.4 DISCUSSION

The reviewed papers show that, mainly due to cost considerations, the quadcopter is the most widely used drone type. It's cheaper, easier to fly, and suitable for small and medium farms. But when the area is large, either multiple quadcopters are flown at once, or drones with a longer range are used.

Furthermore, we have seen from the review that drones are mainly used for agricultural management, especially for identifying diseases. That makes sense because early detection may reduce harm and help farmers save their crops. Drones are used for more than only identifying diseases, despite this being their most common purpose. They are also helpful in mapping, spraying, tracking plant development, and controlling soil moisture. This means that drones are probably used for many different kinds of tasks, possibly all at once. An interesting feature of the research is the different types of methods used; some use simple cameras, while others make use of multispectral sensors, image processing, and artificial intelligence. This could mean that different approaches are used for different problems, or it could mean that researchers are still trying to find the best approaches. In any case, it shows that the area is still evolving and that there isn't yet a single best method.

A challenge that mostly comes up is the lack of technical skills needed to operate drones and how expensive they are. Although this technology has a lot of potential, it may be difficult to deploy in some places, especially in underdeveloped countries.

However, overall, drones continue to show themselves to be solid, practical tools for modern agriculture, but there is still room for improvement, particularly with regard to certain techniques, training, and lowering their cost.

12.5 CONCLUSION

Drones are now important devices used in precision agriculture, which provides farmers new and different ways to manage their crops, control resources, and even increase crop production. Based on the review, we've seen that quadcopters are the most popular drone type because of their stability, being easy to use, and being able to stay in the air for full data collecting. They are mostly used for monitoring plant health, identifying pests and diseases, and helping with spraying, among other things. These help farmers in making better choices while using less fertilizer, water, and pesticides. Despite the high costs, regulatory restrictions, and a lack of technical knowledge in remote areas, the use of drones in agriculture is still expanding.

REFERENCES

Abdulridha, J., Ampatzidis, Y., Qureshi, J., & Roberts, P. (2020). Laboratory and UAV-based identification and classification of tomato yellow leaf curl, bacterial spot, and target spot diseases in tomato utilizing hyperspectral imaging and machine learning. *Remote Sensing*, 12(17), 2732. https://doi.org/10.3390/rs12172732

Akdoğan, C., Özer, T., & Oğuz, Y. (2024). Design and implementation of an AI-controlled spraying drone for agricultural applications using advanced image preprocessing techniques. *Robotic Intelligence and Automation*, 44(1), 131–151.

Albetis, J., Duthoit, S., Guttler, F., Jacquin, A., Goulard, M., Poilvé, H., Féret, J.-B., & Dedieu, G. (2017). Detection of Flavescence dorée grapevine disease using unmanned aerial vehicle (UAV) multispectral imagery. *Remote Sensing*, 9(4), 308. https://doi.org/10.3390/rs9040308

Berner, B., & Chojnacki, J. (2017). Use of drones in crop protection. *Journal of Research and Applications in Agricultural Engineering*, 62(4), 5–9.

Berni, J. A. J., Zarco-Tejada, P. J., Suárez, L., & Fereres, E. (2009). Thermal and narrowband multispectral remote sensing for vegetation monitoring from an unmanned aerial vehicle. *IEEE Transactions on Geoscience and Remote Sensing*, 47(3), 722–738.

Burgos, S., Mota, M., Noll, D., & Cannelle, B. (2015). Use of very high-resolution airborne images to analyse 3D canopy architecture of a vineyard. *The International Archives of the Photogrammetry, Remote Sensing and Spatial Information Sciences*, XL-1/W4, 399–403.

Caballero, D., Calvini, R., & Amigo, J. M. (2019). Hyperspectral imaging in crop fields: Precision agriculture. In F. Marini (Ed.), *Data handling in science and technology: Vol. 32. Chemometrics in food chemistry* (pp. 453–473). Elsevier.

Celen, I. H., Önler, E., & de Lyon, L. B. (2020). Drone technology in precision agriculture. In H. I. Kurt (Ed.), *Academic studies in engineering sciences* (pp. 121–140). Livre de Lyon.

Chin, R., Catal, C., & Kassahun, A. (2023). Plant disease detection using drones in precision agriculture. *Precision Agriculture*, 24(5), 1663–1682.

Daponte, P., De Vito, L., Glielmo, L., Iannelli, L., Liuzza, D., Picariello, F., & Silano, G. (2019). A review on the use of drones for precision agriculture. *IOP Conference Series: Earth and Environmental Science*, 275, 012022. https://doi.org/10.1088/1755-1315/275/1/012022

Duggal, V., Sukhwani, M., Bipin, K., Reddy, G. S., & Krishna, K. M. (2016). Plantation monitoring and yield estimation using autonomous quadcopter for precision agriculture. In *2016 IEEE International Conference on Robotics and Automation (ICRA)* (pp. 5121–5127). IEEE. https://doi.org/10.1109/ICRA.2016.7487716

Garg, P. K. (2022). Characterisation of fixed-wing versus multirotors UAVs/drones. *Journal of Geomatics*, 16(2), 152–159.

Gnädinger, F., & Schmidhalter, U. (2017). Digital counts of maize plants by unmanned aerial vehicles (UAVs). *Remote Sensing*, 9(6), 544. https://doi.org/10.3390/rs9060544

Hassan, S. I., Alam, M. M., Zia, M. Y. I., Rashid, M., Illahi, U., & Su'ud, M. M. (2022). Rice crop counting using aerial imagery and GIS for the assessment of soil health to increase crop yield. *Sensors*, 22(21), 8567. https://doi.org/10.3390/s22218567

Jeong, S., Ko, J., Kim, M., & Kim, J. (2016). Construction of an unmanned aerial vehicle remote sensing system for crop monitoring. *Journal of Applied Remote Sensing*, 10(2), 026027. https://doi.org/10.1117/1.JRS.10.026027

Khose, S., Dhokale, B., & Shashank, S. (2023). The role of precision farming in sustainable agriculture: Advancements and impacts. *International Journal of Environment, Agriculture and Biotechnology*, 8(5), 115–119. https://doi.org/10.22161/ijeab.85.14

Kitano, B. T., Mendes, C. C. T., Geus, A. R., Oliveira, H. C., & Souza, J. R. (2019). Corn plant counting using deep learning and UAV images. *IEEE Geoscience and Remote Sensing Letters*, 16(11), 1674–1678.

Meivel, S., & Maheswari, S. (2021). Remote sensing analysis of agricultural drone. *Journal of the Indian Society of Remote Sensing*, 49(3), 689–701.

Mekonnen, Y., Namuduri, S., Burton, L., Sarwat, A., & Bhansali, S. (2019). Machine learning techniques in wireless sensor network-based precision agriculture. *Journal of the Electrochemical Society*, 167(3), 037522. https://doi.org/10.1149/2.0222003JES

Primicerio, J., Di Gennaro, S. F., Fiorillo, E., Genesio, L., Lugato, E., Matese, A., & Vaccari, F. P. (2012). A flexible unmanned aerial vehicle for precision agriculture. *Precision Agriculture*, 13(4), 517–523.

Santesteban, L. G., Di Gennaro, S. F., Herrero-Langreo, A., Miranda, C., Royo, J. B., & Matese, A. (2017). High-resolution UAV-based thermal imaging to estimate the instantaneous and seasonal variability of plant water status within a vineyard. *Agricultural Water Management*, 183, 49–59. https://doi.org/10.1016/j.agwat.2016.08.026

Shaw, K. K., & Vimalkumar, R. (2020). Design and development of a drone for spraying pesticides, fertilizers and disinfectants. *International Journal of Engineering Research & Technology*, 9(5), 1127–1132. https://doi.org/10.17577/IJERTV9IS050543

Spoorthi, S., Shadaksharappa, B., Suraj, S., & Manasa, V. K. (2017). Freyr drone: Pesticide/fertilizers spraying drone—An agricultural approach. In *2017 2nd International Conference on Computing and Communications Technologies (ICCCT)* (pp. 252–255). IEEE.

Thamm, H. P., Brieger, N., Neitzke, K. P., Meyer, M., Jansen, R., & Mönninghof, M. (2015). SONGBIRD—An innovative UAS combining the advantages of fixed wing and multi rotor UAS. *International Archives of the Photogrammetry, Remote Sensing and Spatial Information Sciences*, XL-1/W4, 345–349. https://doi.org/10.5194/isprsarchives-XL-1-W4-345-2015

Veroustraete, F. (2015). The rise of the drones in agriculture. *EC Agriculture*, 2(2), 325–327.

Vohra, D. S., Garg, P. K., & Ghosh, S. K. (2024). Bringing revolution through quadcopter technology in the field of supplying medicine in the Himalayan regions of Uttarakhand: Case example of Pithoragarh. *Irish Journal of Medical Science*, 193, 265–270. https://doi.org/10.1007/s11845-023-03400-y

Chapter 13

Blockchain governance and legal frameworks for sustainable UAV ecosystems

A global and Indian perspective

Rituraj Sinha, Sheetal Singh, Vijeta Verma, Shashank Solanki, and Vijaishree Dubey Pandey

13.1 INTRODUCTION

13.1.1 UAV governance as a live regulatory problem

There has been an increase in the rate of civilians using Unmanned Aerial Vehicles (UAVs) in many sectors of the economy, such as logistics, last-mile delivery, disaster management and humanitarian aid, environmental surveillance, smart city infrastructure, and aerial surveillance [1]. The rapid implementation of UAV technologies in these fields has fundamentally transformed the nature of airspace operations, turning it into a domain of conventional manned aviation and making it a stratified, data-rich, and highly autonomous ecosystem. Owing to the increasing application of UAVs in the low-altitude airspace and in urban missions, the regulatory questions related to the technology can no longer be viewed as equally hypothetical and speculative assumptions; rather, they have become direct and ongoing governance issues to the aviation regulators and policymakers worldwide.

There have been three chronic pressure areas that have emerged regularly. Firstly, licenses and authorization procedures in most jurisdictions remain slow and bureaucratic in nature, and thus it has adverse impacts on the utility of UAVs that require speed, particularly when responding to emergencies and in times of disaster relief. Second, the compliance is ambiguous due to cross-jurisdiction in law regimes that govern aviation safety, data protection, cybersecurity, and municipal land use because they confuse the operators and the agencies that enforce such laws [2]. Third, an enforcement gap still exists due to the absence of real-time monitoring and due to the absence of reliable and tamper-proof compliance records, especially in an indirect or a transnational activity of UAVs.

The traditional aviation regulation which is founded on a command and control model that was developed in manned aircrafts are ill adapted to address these problems [3]. Such regulatory models presuppose the centralized management of human and rarefied flights, which do not exist in the case of autonomous or semi-autonomous UAV fleets in large-scale flights. Consequently, the regulatory design and real operationality become increasingly distanced, and this necessitates a fresh perspective in the devices of

DOI: 10.1201/9781003731580-13

governance that can facilitate safety, accountability, and sustainability in the current UAV ecosystem.

13.1.2 Practical problem statement (experiential focus)

The operations of UAV are also complicated by the fact that the zones of the legal regime are more divided, primarily, the border of the aviation law, data protection law, and digital governance frameworks. The primary aviation regulations are safety, airworthiness, and airspace management; the data protection laws are also applied to regulate the collection, processing, and storage of personal and sensitive information collected by UAV sensors [4]. Such digital governance standards as cybersecurity and digital identity administration are complementary to each other but are not sufficiently reflected in the law. Practical challenges of incarceration of regulations on a political level are serious.

The compliance in the UAV operator's case can be viewed to include the working of several regulatory bodies, technical requirements, and on numerous occasions, increased costs, time wastages, and law uncertainty. The issue of keeping track of UAVs in real time and post-flight non-compliance is a challenge to aviation regulators; the lack of an easy method of distributing the responsibility when it comes to autonomous decision-making or system failure is also a problem. The local and city governments, taking care of the security of citizens and the urban planning, usually do not have access to the flight data and the possibility to utilize the mechanism that could be implemented to manage the low-altitude UAV operations within their jurisdiction. Similarly, structural bottlenecks are encountered in the emergency response agencies since the rapid implementation of the UAV during the occurrence of natural disasters or a pandemic outbreak is required due to the necessity to utilize the available authorization structures in dynamic and time-sensitive scenarios [5]. These real-world incompetencies are signs of a natural lack of a fit between regulatory will and operational reality. This kind of inadequacy of a well-developed governance infrastructure with which the regulation adherence, data, and transparency of operations accountability can be coordinated undermine the assurance of the law and citizen trust. To fill this dark side, there is the need to shift toward a more technologically enabled governance system than the normative regulation system that cannot work in the real-world UAV operations.

13.1.3 Original contribution of the chapter

It is on the basis of this that this chapter constructs a legal, blockchain-oriented system of governance of UAV ecosystems that alters the disjunction between the regulatory theory and practice. The conceptualized chapter views blockchain as a regulatory framework and not a technological invention relying upon the comparative aviation law and digital governance framework and therefore is capable of installing compliance, accountability, and traceability as a direct element of UAV operations. The chapter offers a convenient legal framework, which is innovative and regulatory integrity-balancing, in mapping out legal demands by practical regulatory procedures (such as licensing, flight authorization, monitoring, and post-incident auditing).

What makes the provided contribution special is that it provides a practical legal investigation that puts the blockchain governance into perspective of the already accepted international aviation standards and domestic laws, namely, the existing alterations in the UAV regulation in India. Rather than proposing abstract changes, the chapter evaluates how the blockchain-based governance mechanism can be operational in the context of the legal constraints of the time in order to multiply the outcomes of the regulatory efficacy and sustainability [6].

13.1.4 Methodology

It also has an analytical methodology of a comparative doctrinal legal approach, tools of international aviation, its territorial regulatory framework, and national laws that govern the behaviors of UAVs. This theoretical review is complemented by a case-based regulatory evaluation, which is based on judicial decisions, regulatory experiences, and policy events that are applied in the UAV regulation and data security. Other than that, the chapter employs a policy design and compliance-mapping strategy that identifies the areas in which the blockchain technology can be integrated as being a part of the regulation processes without posing a threat to the constitutional or statutory protections. The course of the approach also informs the analysis of the practitioner-oriented legal reasoning; it is concerned with the regulatory feasibility, enforceability, and institutional capacity as given in Table 13.1.

13.2 OPERATIONAL ANATOMY OF UAV ECOSYSTEMS (PRACTICE-ORIENTED)

13.2.1 Stakeholders in real UAV operations

The legal and regulatory frameworks of the ecosystem stakeholders are complex and interdependent but different, which characterizes the UAV ecosystems. The key players in the management of the airspace, safety, and licensing are the aviation authorities, such as the Directorate General of Civil Aviation (DGCA), the Federal Aviation Administration (FAA), and the European Union Aviation Safety Agency (EASA). Their regulatory tasks are expanding to a new level where the traditional aviation concerns are accompanied by the data on the digital identity management of remote identities and real-time monitoring of UAV operations [7].

Manufacturers of UAV play a vital role in ensuring airworthiness of aircrafts, integrity of programs, and conformance to technical specifications, but operations and service providers have direct liability in regard with legal deployment, compliance to flight authorizations, and safety of operations.

Table 13.1 Blockchain functions mapped to key UAV regulatory challenges

Regulatory challenge	*Existing gap*	*Blockchain intervention*	*Legal/regulatory relevance*
UAV licensing and registration	Delays and lack of transparency	Smart contract-based automated licensing	Drone Rules, 2021; FAA Part 107; EASA UAS Regulation
Flight authorization (pre-flight)	Slow, centralized approvals	Automated permission validation	Digital Sky Platform; ICAO RPAS Guidelines
Post-incident investigation	Risk of data loss or manipulation	Immutable flight and telemetry logs	Aviation safety law; evidentiary standards
Liability determination	Difficulty attributing fault in autonomous flights	Time-stamped operational records	Tort law; product liability principles
Multi-agency regulatory oversight	Fragmented enforcement and data silos	Shared, permissioned regulatory ledger	Administrative coordination frameworks
Data governance and privacy	Unclear access and compliance mechanisms	Controlled, auditable data access	GDPR; DPDP Act, 2023; *Puttaswamy* (2017)
Emergency UAV deployments	Regulatory rigidity in crises	Pre-authorized smart contracts	Disaster Management Act, 2005
Sustainability and ESG reporting	Limited verifiable operational data	Immutable environmental data records	SDGs 11 & 13; environmental governance

Moreover, the data processors and local authorities have turned out to be major stakeholders because UAV activities are data-driven, and the impact of low-altitude drone-based operations is local. The existence of such plurality of actors necessitates the existence of governance structures that would aid in aspects of coordination, accountability, and transfer of information across institutional boundaries.

13.2.2 Regulatory touchpoints across the UAV lifecycle

Operationally and legally, UAV governance can be conceptualized as a series of regulatory contacts during the entire lifecycle of a UAV operation. The legal personhood of the UAV and its operator, the basis of accountability, is determined during the registration step. Flight authorization systems specify the time, location, and conditions of flight allowed by the UAVs, which typically involve airspace boundaries and safety aspects. Real-time performance and remote identification are required in the in-flight phase to enforce and prevent collisions, particularly in the congested or sensitive airspace [8].

The incident reporting and investigation systems are important post-operation in order to address accidents, data breaches, or regulatory breaches. Finally, the post-flight audit process and the establishment of the liabilities are founded on the adequate and verifiable documentation of the operations in order to resolve conflicts and to apply fines. In other words, the absence of one system through which to interlink these points of regulation makes it much easier to find fragmentation and inefficiencies in the implementation. The analytical basis of this lifecycle-based perspective of UAV governance can be applied to identify how the blockchain-based systems can be applied to integrate compliance and accountability throughout the entire phases of UAV operations.

13.3 BLOCKCHAIN AS AN APPLIED REGULATORY INFRASTRUCTURE

The UAV ecosystem policy is increasingly necessitating regulatory actions that are not only normative but also efficient in operations as well [9]. The current method of aviation regulation to a large extent centrally licensing-based, ex post-regulation, and manual compliance-checking, struggles to keep up with the scope, autonomy, and data-intensive nature of contemporary UAV operations. In this connection, blockchain technology is not just an auxiliary one but more a form of applied regulatory infrastructure, which has the potential to directly introduce legal compliance into the working environment of the UAV ecosystems [10].

The blockchain systems are smart contract-programmable, decentralized, and immutable, as compared to the conventional digital databases. Such properties then allow the imposition of regulatory requirements—e.g., licensing, permission to fly, and maintenance requirement—ex ante and on a continuous basis, in contrast to being imposed ex post, through enforcement actions. In turn, blockchain enables transitioning to active regulation of the aviation sector to the model of embedded compliance, where a legal provision is substantiated by default within the technological framework governing the activity of UAVs.

13.3.1 Why blockchain fits UAV governance (operational rationale)

To ensure a high degree of reliability of compliance with the highly decentralized ecosystem made up of multiple operators, platforms, and regulatory bodies is one of the most significant issues of UAV regulation. This issue can be resolved using blockchain in this aspect since it possesses three, interlinked, operational characteristics [11].

Firstly, raw compliance logs provide a dynamic free-hand record of UAV-related activities, e.g., registration, flight approvals, telemetry data, and incident documents. There is also the legal benefit of these records, which can never be changed, that makes evidentiary integrity of post-incident investigations, liability adjudication, and regulatory audits [12]. The logs based on blockchains can enhance transparency and credibility, unlike the traditional logs, which can be altered or published at random times, to the regulators as well as the concerned third parties.

Second, smart contracts can provide automated permissions that can be put in coded regulatory approvals. Using the example of an issue of flight permissions, it can be offered on a conditional basis, provided that it is in compliance with the airspace restrictions, operator qualification, and time limitations. In case all these conditions are satisfied, then you are not delayed to authorize. The model is highly congruent with No Permission, No Takeoff (NPNT) model of the Drone Rules, 2021, and envisions how Drone Rules can be pragmatized programmably as opposed to the manual approvals [13].

Third, the regulators should be audited in a decentralized manner, enabling a common verifiable compliance ledger to be held by various authorities such as the aviation regulators, law enforcement agencies, and environmental agencies. This is of greater significance when the institutional supervision of UAVs is shattered. Blockchain thus enables coordinated oversight, bypassing the need to create one dominant enforcer, hence adhering to the principles of administrative decentralization and improving the coherence of regulatory practices [14].

13.3.2 Practical use-cases already emerging

Applying blockchain to the control of UAVs is not just a theory anymore. The number of practical use cases is already quite large in most jurisdictions, which is an indication it can serve as a valuable means of regulation.

One such application is Remote Identification (Remote ID) authentication because with blockchain-based identity registries, it is possible to verify UAVs and operators on a real-time basis. By linking individual digital identities to non-mutable registers, regulators will be able to get traceability without necessarily relying on continuously centralized data sets. This is particularly a point of concern in regards to unauthorized or malicious drone operation in sensitive airspace [15].

The second application is flight permission ledgers, which are records of authorizations, geofencing restrictions, and time-limited approvals. Such ledgers would leave a distinct trail of compliance, and regulators could at any one point in time check that the UAV flight was in compliance. These systems reduce disputes on ex post compliance and legal certainty for the operators [16].

Third, the record keeping through blockchain makes it possible to maintain and monitor airworthiness, and thus, preserve the lifecycle data, such as repairs, all checks, and component modification on a blockchain. This is especially pronounced as the UAVs operations are augmented and autonomous flight operations are augmented [17]. Permanent records of maintenance are legal in that they are used to spread the liability among the operators, the manufacturers, and the software vendors in the event of failure or accidents of the system (Table 13.1).

13.4 INTERNATIONAL LEGAL ARCHITECTURE: LIMITS IN PRACTICE

Although UAV ecosystems have evolved technologically, the international law applied in aviation has not developed much since its application was based on the principles of manned aircraft. Though these frameworks offer the much-needed normative underpinnings, they have serious limitations when used in the case of the decentralized, blockchain-based UAV operations [18].

13.4.1 Chicago convention, 1944 – operational shortcomings

The Chicago Convention of 1944 (Convention on International Civil Aviation) is still the foundation as far as international aviation law is concerned. Article 8 of the Convention covers the concept of pilotless aircraft and states that the pilotless aircraft cannot be allowed to fly over the territory of a contracting state without special authorization [19]. Although this provision gives state sovereignty on the airspace, it provides few indications on how the authorization, monitoring, and enforcement can be implemented in technologically challenging UAV settings.

Interestingly, the Chicago Convention does not consider computerized enforcement means, real-time compliance verification, and automatic authorization systems. Consequently, states are deprived of a common global framework that can balance blockchain-based governance by coming up with domestic regulatory solutions [20]. This brings about a legal doubt in cross-border UAV operations, especially where the flight data, control systems, and blockchain nodes can be spread over different jurisdictions. Additionally, the very inflexible territorial notion of the sovereignty that is entrenched in the Convention makes the utilization of decentralized ledgers, that are otherwise ill-suited to their country borders, rather difficult. Whether the applicable law, jurisdiction, and state responsibility are upheld in the case UAV activity falls under the jurisdiction of transnational digital infrastructures, the questions about these issues remain unanswered [21].

13.4.2 ICAO RPAS guidance – normative but non-binding

Following the increasing application of remotely piloted aircraft systems (RPAS), the International Civil Aviation Organization (ICAO) has published manuals and guidance materials that should facilitate the formulation of domestic UAV regulations by States. Although these tools do not ignore the importance of technological integration and risk-based regulation, they are soft law and have no binding effect [22].

One of the major weaknesses of ICAO RPAS guidance is that it does not offer blockchain-related standards in the area of licensing, identity management, or compliance logging. Therefore, States that implement blockchain-based UAV infrastructures do it in regulatory seclusion, without internationally accepted standards of interoperability, data control, or legal liability. Such normative gap has a practical implication [23]. In the absence of harmonized international standards, the UAV governance based on blockchain will contribute to the regulation fragmentation instead of alleviating it. Mutual recognition of digital credentials and compliance records is also constrained by the absence of binding ICAO norms and impedes the possibility of an unproblematic cross-border UAV operation.

13.4.3 Analytical significance

The combination of these restrictions underscores a severe lack of alignment between the international aviation law as it is and technological regulation as it is becoming. Although blockchain can be used to address regulatory issues in the UAV ecosystem in a practical way, to successfully implement it, the legal adjustment at the global level is needed [24]. This stress highlights the necessity of a remodeled legal framework one that maintains elements of principle of sovereignty and safety whilst facilitating technologically intrinsic mechanisms of compliance that would meet the requirements of sustainable, decentralized UAV operation [25].

13.5 COMPARATIVE REGULATORY PRACTICE: EUROPEAN UNION, UNITED STATES, AND INDIA

The control of UAVs indicates more general regulatory philosophies embraced by various jurisdictions on striking a balance between innovation, safety, privacy, and public interest [26]. The analysis of the European Union, the United States, and India shows that they have different views on UAV regulation, which is why each of them has a different level of preparedness to integrate with blockchain-based governance.

13.5.1 European union (EASA): risk-based but data-heavy

The European Union has implemented a risk-based regulatory framework on UAV operations by the European Union Aviation Safety Agency (EASA). The operations of UAVs are classified into open, specific, and certified estimates which are based on the risk of operation and the compliance requirements are also based on these risk categories. With this method, the regulatory proportionality and flexibility can be provided especially in the case with low-risk civilian drone missions [27].

Nevertheless, although the EASA framework is technologically advanced, its applicability in the context of blockchain-based UAV ecosystems is limited by the fact that the EU has an advanced data protection system, which is the General Data Protection Regulation (GDPR). The work of UAVs produces large volumes of personal and non-personal information, such as images, geolocation, and behavioral information, in most cases, falling under the broad definition of personal data in the GDPR [28]. The immutability that is the key property of blockchain poses an inherent conflict with the postulates of GDPR, including data minimization, purpose limitation, and most importantly, the right to erasure (Article 17). In terms of practical compliance, there is more regulatory complexity in trying to implement decentralized ledger technologies in UAV governance systems, by the operators. Although such workarounds as off-chain storage, hashed identifiers, and permissioned blockchains are being brought up more frequently, they tend to undermine the very decentralization advantages that blockchain is supposed to provide. Consequently, the EU model, despite its normative strength, is cautious and comparatively restrictive to the overall implementation of blockchain in the governance of UAVs.

13.5.2 United States (FAA): innovation-first but fragmented

The United States takes an innovation-first regulatory stance spearheaded by the Federal Aviation Administration (FAA), especially by Part 107 of the Federal Aviation Regulations. The FAA model is based on the priorities of commercial scalability, experimental deployment, and involvement of the private sector, and, therefore, the US is an excellent location to conduct UAVs innovation and testing. However, this liberal policy is also associated with high regulatory fragmentation. There are often overlaps between the federal regulations of UAVs and state and local laws of privacy, land use, and surveillance. In judicial interventions, City of Newton has raised conflicts between federal pre-emption and local regulatory autonomy, which has put legal confusion on UAV operators [29].

In the context of the blockchain governance, the US system does not have a single data protection law similar to the GDPR. Although this gives it operational flexibility, it also has the effect of creating an uneven enforcement standard, ineffective accountability mechanisms, and minimal automation of the regulatory process [30]. Lack of a unified federal privacy policy makes it difficult to implement the blockchain-based UAV compliance frameworks, especially the cross-border or interstate one. In this way, the US system is institutionally loose and light on enforcement but is technically receptive.

13.5.3 India: digital sky as a De Facto GovTech experiment

The Drone Rules, 2021, gives a unique regulatory model in India, where the UAV operations were liberalized considerably, and they were integrated into a centralized system of digital governance. Digital Sky platform is a one-window registration/licensing/airspace authorization/compliance monitoring system. The NPNT system is an example of applying the concept of real-time regulation enforcement by implementing the legal compliance in the UAV operations technologically [31].

Digital Sky is a proto-blockchain platform, and it provides traceability, authentication, and auditability, even though in a centralized form. This places India in a relatively good standing with regard to integration of blockchains in the future, especially aspects like automated licensing, compliance logs that cannot be altered, and decentralized supervision [32]. Nevertheless, there is still a question of scalability. As the UAV functions in the fields of logistics, disaster management, and surveillance of cities, a strictly centralized system can face constraints on the load of the system, its coordination between agencies, and cross-border interoperability. The future sustainability of the UAV governance infrastructure in India might be limited without gradual decentralization which can be achieved by allowing permissioned blockchain architecture (Table 13.2).

13.6 CONSTITUTIONAL AND DATA GOVERNANCE CONSTRAINTS (INDIA-FOCUSED)

India is perceived to be very innovative in its regulation of the organization of the UAVs; however, the privacy of the personal data and the constitutional limits constrain the development and deployment of the blockchain-based UAV systems to a significant extent.

13.6.1 Informational privacy and UAV-generated data

The constitutional foundation of Indian data regulation is based on the decision of the Supreme Court on Justice K.S. Puttaswamy v. Union of India (2017). The right to privacy that was set forth in the Article 21 of the Constitution was established in the Union of India (2017) and is considered a fundamental right. The Court mentioned the proportionality test, which contained three prongs that comprise legality, legitimate state purpose, and necessity that is accompanied by proportional safeguards [33].

In the case of the application to the use of the UAVs, especially the aerial surveillance missions or the data-intensive operations in the logistical operations, this framework demands very high standard of scrutiny of the data collection, data storage, and data processing practices. Governmental approaches of UAVs, which are built on blockchains and value permanence and traceability, should be optimized to ensure that the intrusion into informational privacy is not excessive. To be acceptable to the constitutional review, any comprehensive use of blockchain-recorded UAV information would have to demonstrate at least a deficiency of data retention, intentional use, and effective controls.

Table 13.2 Comparative regulatory readiness for blockchain–UAV integration

Parameter	*EU (EASA/GDPR)*	*USA (FAA/federal & state law)*	*India (DGCA/drone rules 2021)*	*Practical implications*
Digital infrastructure	High: Advanced UTM and operator databases	Medium: Fragmented federal-state systems	High: Digital Sky Platform for real-time monitoring and compliance	EU and India can support blockchain integration at scale
Blockchain compatibility	Low: GDPR constraints on immutable data	Medium: Flexible but no uniform blockchain policy	High: Policy support for blockchain-enabled governance	India can pilot smart contract-based licensing; EU needs privacy-compliant design
Enforcement automation	Medium: Risk-based authorizations partially automated	Low: Mostly post-flight compliance	Medium: NPNT-linked automated approvals and monitoring	India and EU can improve real-time enforcement; US lags
Licensing & registration	High: Centralized EU database	Medium: FAA registry, state-level delays	High: NPNT-linked automatic approvals	Blockchain can streamline licensing in EU and India
Cross-border operations	Medium: ICAO alignment exists, GDPR issues	Low: Limited mutual recognition of UAV credentials	Medium: Digital Sky IDs evolving for international use	Blockchain-enabled credentialing needed for compliance
Data privacy and security	Low: GDPR restricts immutable personal data storage	Medium: State-level privacy laws; no uniform federal law	Medium-High: DPDP Act allows permissioned ledgers	Privacy-preserving blockchain needed, especially in EU
Sustainability/ESG	Medium: Emerging incentives for low-carbon UAVs	Low: Minimal ESG integration	High: SDG-aligned UAV initiatives possible	Blockchain can track carbon footprint and ESG metrics

13.6.2 Digital Personal Data Protection Act, 2023

One of the largest steps toward the statutory data regulation in India is the implementation of the Digital Personal Data Protection Act, 2023 (DPDP Act). Some of the principles that the Act emphasizes include the restriction of purposes, minimization of data, and consent-based processing and offer the data principals a binding right. However, there are structural issues of incorporating blockchain within the DPDP Act [34]. The indelible registries also go against the requirement to delete or correct individual data, which leaves the issue of the legality of decentralized systems incompletely resolved. The Act also assumes consent architectures that are effectively revocable and dynamically governed and are difficult to implement in a blockchain environment.

With UAV governance, it would need a mixed-knowledge-law relation, wherein sensitive personal information is stored out-of-chain, with blockchain used to verify, store metadata about, and prove regulatory compliance. In the absence of such design decisions, the UAV governance through blockchains will be threatened by the statutory non-compliance with its operation despite the advantages of blockchain use [35].

13.7 CASE-BASED PRACTICAL LEGAL ANALYSIS

In this part, the theoretical discussion of the blockchain-driven UAV governance seems to be grounded on the real judicial logic and rule-making. It demonstrates how courts and legal systems are responding to the emerging challenges posed by the operations of UAVs, decentralized compliance, and data-intensive aerial technologies, according to the analysis of the previously existing case law and jurisprudential principles.

13.7.1 International case example: *Singer v. City of Newton* (United States)

The decision in Singer v. City of Newton is one of the most educative judicial contacts with the tension of regulation of civilian UAV operations in the US. The case began after the City of Newton, Massachusetts, enacted local ordinances, including the restrictions on the drone flights that included the need to register the drone flight with the local authorities, the impositions on aerial photographing of privately owned lands, and the impositions of the minimum altitude of the drone flight without prior permission of the land owners [36]. The areas that were objected on were that the stipulations were inconsistent with the federal aviation regulatory system under Federal Aviation Administration (FAA). The legal issue that the court was to determine was the federal pre-emption and local regulatory authority. The FAA argued that the navigation of the airspace, aircraft safety standards, and operating authority were all within the federal jurisdiction. The court agreed on most of the points and it was believed that local laws which attempted to control the height of the flight, authorization of operations or the air that was navigable, encroached an area that was under federal law [37]. Meanwhile, the court granted that the municipalities retain a limited scope of power to manage land use, privacy, and local law enforcement matters, but they cannot in fact airspace regulation [38].

Singer v. City of Newton gains a few vital points in the context of the blockchain-based UAV governance. First, it brings out the importance of uniformity and interoperability of regulations, particularly when the UAV activity crosses into the local borders. Decentralized compliance systems (notably the distributed flight authorization ledger, remote identification logs, etc., based on blockchains) must be designed in such a manner that they are federally harmonized and not a local patchwork. Second, the case throws light on the value of technological compliance solutions that are able to integrate the federal requirements into working processes. In the broader sense, by using the example, Singer v. City of Newton demonstrates that being a decentralized compliance system does not mean that legal hierarchy is nonexistent [39]. On the contrary, such systems are to be structured in a manner that they do not interfere with constitutional allocations of regulatory authority. However, blockchain governance must not be an alternative to any legal frameworks but instead, it should be a compliance infrastructure overridden on them.

13.7.2 Indian jurisprudential application: *Justice K.S. Puttaswamy v. Union of India* applied to UAV surveillance

However, there are no attempts to use a judicial precedent to solve the particular problem of UAV surveillance specifically in the Indian setting, as the constitutional principles are reflected in the case Justice K.S. Puttaswamy v. The article by Union of India provides a good jurisprudential base to the discussion of UAV-based data processing and blockchain-based logging solutions. The establishment of the Right to Privacy as a key right by the Supreme Court in article 21 of the Constitution has extremely far reaching implications on the aerial surveillance, data collection, and round-the-clock surveillance offered by UAV technologies [40].

By using Puttaswamy v. Union of India in UAV surveillance, this denotes three dimensions that are constitutionally pertinent. First, UAVs possessing high-resolution images, sensors, and AI analytics are capable of collecting the details of personal information, and in most instances, without the information or authorization of the people involved. Second, in the case when such information is uploaded permanently in blockchain registries and is intended to serve as a compliance, audit, or evidentiary tool, the issues of informational self-determination and data minimization are brought up. Third, the proportionality test that was embraced in Puttaswamy v. Union of India that includes legality, legitimate aim, necessity, and proportionality would place on UAV surveillance, including UAV surveillance that is technologically efficient and capable of being effectively refined, a burden of proving expressly drafted statutory power and narrowly focused purpose. The impossibility to change distributed ledgers may be unsuitable with other rights, such as the right to correction, erasure, or limitation of use, which are partially reflected in the Digital Personal Data Protection Act, 2023. A hypothetical judicial application of Puttaswamy to stop blocking chain-logged UAV data would most likely entail courts discovering some manner of drawing the distinction between personal data stored on-chain and hashed data structures and reference-based data structures, which would preserve auditability but not disclose identifiable information. This reading would back the privacy-by-design approach of regulating UAVs, rather than a blanket rejection of blockchain technologies [41].

Thus, the constitutional jurisprudence of India does not prohibit the regulation of the blockchain-enhanced UAV as long as it is followed by the principles of proportionality, responsibility, and the law [42].

13.8 DETAILED LEGAL–OPERATIONAL CASE STUDY

13.8.1 Case study: blockchain-enabled UAV deployment in disaster management

In this part, a practical scenario will be used as to why blockchain governance is relevant to the UAV ecosystem in a real-life example with reference to the introduction of UAVs in the event of a significant natural disaster such as a flood or an earthquake. The disaster response context is an interesting case to use as it is a combination of urgency, the coordination of multiple agencies, the sensitivity of data, and a higher level of interest among people [43].

13.8.2 Facts (realistic regulatory scenario)

Consider the example of an instance when the UAVs are being used by the government agencies to survey a massive flood that has struck various districts. UAVs are used in real-time mapping, searching lost individuals, providing medical services, and examining damaged infrastructure. Depending on the number of stakeholders involved, they include disaster management officials, aviation regulators, local governments, and individual service companies for UAVs. The system relies on a blockchain to record flight authorizations, UAV flight paths, sensor data tagging, and an immutable audit trail, which may be viewed by approved agencies [44].

13.8.3 Legal issues

Such a state of affairs raises several interconnected legal issues. The first one is the emergency exceptions to privacy: despite the possibility of disaster response to become the reason to temporarily suspend the consent and the notice requirement, the limits of the exceptions along with the time frame must be defined by the law [45]. The second issue is linked to the responsibility of independent or semi-autonomous decision-making, particularly, with regards to UAVs relying on AI-based navigation or prioritization routines. The decision-making sharing between hardware, software, and data inputs is rendered hard to define responsibility in case of errors that can occur during the targeting or collisions. The third issue is the retention and auditability of data, especially when the records in blockchain storing operational data are not subjected to the emergency in question and they are not necessarily aligned with the values of data minimization.

13.8.4 Legal analysis

Even under the Constitution, proportionality and necessity have to be achieved in deployment of the UAVs during an emergency. The case of a disaster is also likely to ensure that the courts uphold few breaches of privacy, provided there is some statutory support, time limitation, as well as after-event responsibility. The national drone needs of aviation safety exist only with expedited approval or exemption on condition. Particularly, blockchain governance is relevant in this respect, as it enables ex ante compliance (through automated authorizations) and ex post accountability (through immutable logs) [46].

The data governance analysis further shows that blockchain is not a mandatory storage of personal information, as it can be employed as a verification layer, where hashes, time stamps, and authorization logs are stored, but the sensitive data are not stored on the blockchain but in an off-chain and controlled environment. These architectures are a compromise between disaster-response and compliance efficiency and legal regulations.

13.8.5 Practical outcomes

Practically, UAV management of disasters with the help of blockchain can introduce several measurable advantages. The pre-programmed emergency procedures ensure that the operations of receiving regulatory approvals proceed much faster. The responsibility is also enhanced because all the decisions made regarding the operation can all be traced, and this assists in reducing the post-incident turmoil within the agencies or the contractors. Finally, an institution is able to learn and improve its policies through verifiable records of the operations in case of any future crisis.

13.9 SUSTAINABILITY AND ESG INTEGRATION (APPLIED CONTRIBUTION)

Sustainability issues are no longer the peripheral component of the regulation of technology but the center of norms. Sustainability—the UAV ecosystem context is no longer a phenomenon that is confined to the environmental debate, but it has emerged as a parameter that is now legally significant in the context of rules design, compliance requirements, and social responsibility. Blockchain-based governance systems offer a unique opportunity to implement the principles of Environmental, Social, and Governance (ESG) in the extreme realms of the UAV regulation and perceive sustainability not as a policy objective but as a particular and practical regulatory outcome in the field of operations [47].

13.9.1 UAVs as low-carbon logistics tools: empirical policy reasoning

UAVs, particularly electrically powered drones, are one of the solutions that are viewed as low-carbon alternatives to the conventional logistics and transport systems. UAVs demonstrated that they can be used to reduce fuel use, traffic congestion, and the subsequent greenhouse gas emissions because of their application in last-mile delivery, delivery of medical supplies, disaster recovery, and infrastructure inspection. Regulatively speaking, these environmental benefits directly relate to State obligations to international environmental law, e.g., in the Paris Agreement, and the implementation of the United Nations Sustainable Development Goals (SDGs), e.g., SDGs 9 (Industry, Innovation and Infrastructure), SDGs 11 (Sustainable Cities and Communities), and SDGs 13 (Climate Action).

The sustainability advantages of UAVs must be proven empirically, however, so that regulatory incentives, exemptions, or special treatment can be informed. The blockchain-based governance makes it possible to substantiate it with the data on the verifiability of operations, including the frequency of flights, the energy consumed, the delivery efficiency, and the route efficiency. These unalterable records of data legitimize evidence-based policy-making, and they will allow regulators to understand the actual effect of UAV activity on the reduction of emissions or the yielding to the cost of the environment. In this manner, blockchain will allow sustainability appraisals that are empirically driven, and the legitimacy of the environmental policy intervention by UAVs will be strengthened.

13.9.2 Blockchain-enabled resource optimization and airspace efficiency

In addition to the decrease in emissions, the notion of sustainability in the process of governing UAVs is directly related to the resources and system optimization. The existing regulatory frameworks are characterized by overlapping of the compliance processes, incompatibility of the data storage, and absence of coordination of the aviation authorities, law enforcement agencies, and local governments [48]. Such inefficiencies not only increase the administrative burden but also derail sustainable deployment, increasing the cost of operation and regulatory drag.

One of the solutions to these issues is blockchain-based shared ledgers, where it is possible to access authenticated regulatory data in real-time and in an interoperable manner. During the exchange of licensing documents, flight approvals, compliance records, and incident reports, it is possible to easily exchange them between the authorized parties without having to go through the verification process time and again. This results in the minimization of duplication, coherence of regulations, and enhancing airspace management since it makes it possible to predict traffic and avoid congestion. This optimization improves governance accountability but indirectly contributes to environmental sustainability because the allocation of resources becomes efficient and the wasting of systems is less systematic.

13.10 SUGGESTED TWO-LEVEL GOVERNANCE ARCHITECTURE IN UAV ECOSYSTEMS THAT USE BLOCKCHAINS

With the advent of blockchain technologies in the sphere of logistics, disaster prevention, urban mobility, and environmental monitoring, the application of Unmanned Aerial Vehicles (UAVs) is growing, which promotes new governance issues. The traditional aviation regulatory frameworks prove ineffective to handle the operational complexity, data volume, and transnational characteristics of UAV systems due to the fact that they were mostly implemented to address manned aircraft. In order to give solutions to these challenges, this chapter proposes a two-legged system of governance that balances between, on the one hand, the national regulation and, on the other hand, the international harmonization. The model is devised to provide a frame of reference regarding the manner of possessing clear, responsible, and workable UAV operations and facilitating operational acculturation amid regulators, operators, and technology developer.

13.10.1 Tier I: national blockchain-enabled UAV governance

The lower level is the national level in which confers the United States of America the paramount control to the functioning of UAVs within its airspace. The level is concerned with how the blockchain infrastructure will be implemented in the local regulatory frameworks to ensure that the process of licensing, monitoring, and sharing of the liability becomes more efficient, transparent, and legally certifiable.

13.10.1.1 Licensing and permit issuance

One of the most critical matters in the regulation of the UAVs is the manual licensing of the UAVs that is usually typified by delays, lack of transparency, and discretionary interpretation. The issuance of licenses and operation permits by automated procedures may be achieved using the smart contracts in the blockchain networks in accordance to a set of compliance standards. This reduces the administrative bottlenecks and enhances transparency in the procedures and limits the administrative discretion to control. The efficient manner of application of this concept is the Indian Digital Sky solution, the centralized digital flight approval, and licensing system which can be implemented as a smart contract through a blockchain to provide a possibility to make it even more efficient and transparent.

13.10.1.2 Monitoring and compliance oversight

The safety and regulatory control also requires real-time monitoring of the UAV operations. Blockchain-based Remote Identification (Remote ID) systems have the potential to produce the immutable flight logs and telemetry data that are capable of monitoring the UAVs whereabouts any time. Such a system offers convenience of avoiding enforcement, investigation of the incident, including ensuring that the information of operational data is unalterable and could be heard in a court of law. It is particularly crucial when autonomous or semi-autonomous UAVs are concerned, and the control is somehow left in the hands of people.

13.10.1.3 Liability and risk management

The passing of the responsibility in the UAV operation and autonomous flights in specific cases is a controversial matter. The blockchain technology enables the records involving the operator actions, software interventions, and manufacturer contributions to be made, specifically in time, which enables to distribute liability evidently. It would be applicable in the event of litigious conflicts or insurance and this would be in compliance with tort and product liability and also reduce the uncertainty in regard to litigation.

13.10.1.4 Operational scalability

The point mainly is that at the national level, it is possible to have digital governance because the centralized Digital Sky platform demonstrates. Further improvement of the level of scalability is also possible due to the use of blockchain that will allow the regulatory systems to manage the increasing number of UAVs, operators, and complex operating conditions. This is the automation, transparency, and record keeping that cannot be altered to boost the efficiency at which the operations are carried out and fosters the trust in the stakeholders.

13.10.2 Tier II: international harmonization layer

The second tier is related to the transnational nature of the UAV operations and the need of the regulatory interoperability. The former tier will ensure national regulation of the industry, and the latter will permit mutual recognition of the UAV qualifications, compliance credentials, and operational licenses across jurisdictions. This layer is concerned with harmonization, not uniformity in order that as states retain their sovereign powers, they are still able to use the international UAVs, such as cross-border logistics, disaster management, and urban air mobility.

13.10.2.1 Cross-border regulatory certainty

The UAVs cross-border operation has to possess a legal framework of licensing, adherence to operations, and data management. Blockchain-based credentialing will allow ensuring instant verification of UAV operators and aircrafts and reducing delays and controversies. Interoperable blockchain environments can also enable real-time compliance across jurisdictions in the nation, and this would assist in minimizing the possibility of differing regulations and operational variances [49].

13.10.2.2 Role of international institutions

It is the international organizations, in particular, the International Civil Aviation Organization (ICAO) that is positioned to devise model standards of the blockchain-based governance of UAVs. These standards could advise the states on the development of the compatible regulatory structures, the promotion of safe and sustainable UAV operations, and the aspect of domestic legal customs and priority spheres of the policy agenda. This form of harmonization is compatible with international standards in Chicago Convention, 1944, and ICAO RPAS guidelines, which will provide the format of internationally integrated UAV operations.

13.10.2.3 Legal and policy implications

The liability, data protection, and operational compliance are some of the major implications that the harmonization layer has. The standard blockchain architecture can facilitate the cross-compatibility of digital credentials and the implementation of the cross-border and compatibility with the legal framework at the national scale (such as GDPR in the EU and DPDP Act, 2023, in India). The harmonized systems also enable tracking of the sustainability measures and compliance with the ESG which are under the SDGs of cross-border activities.

13.11 POLICY RECOMMENDATIONS: PRACTITIONER-ORIENTED GUIDANCE

The two-tiered governance model will require some policy recommendations to be provided to the legislators, regulators, and other individuals interested in the industry. These suggestions are expected to interpret the governance theoretical constructions into real and action plans.

13.11.1 Recommendations for legislators

- Adopt Technology-Neutral Frameworks:
  Priorities of the technologies in legislation should be replaced by regulatory interests like safety, accountability, and privacy. This provides the adaptability and evolvability of legislative systems to the change in technology. In the example, the laws may require compliance systems be blockchain ready, but without protocol given in it, industry and regulators can choose the most optimal.
- Encourage Legal Innovation:
  To make sure that the regulation is not dogmatic and unaccommodative of innovations, regulators should encourage regulatory sandboxes and trial projects so that new models of governance, especially blockchain-powered UAVs, can be tested.

13.11.2 Recommendations for regulators

- Establish Sandbox Environments:
  The regulatory agencies ought to have manageable conditions related to tests in which the use of blockchain to govern UAV is tested, such as coordination of cross agencies, privacy, and real-time monitoring.
- Leverage Smart Compliance Audits:
  Audit trails: In the event that an audit trail is maintained in compliance with blockchain, it enables regulators to verify compliance automation, reduce administrative burden, and enforcement to become efficient.
- Enhance Multi-Agency Oversight:
  Shared permissioned ledgers may be useful in coordinated surveillance between aviation authorities, municipalities, disaster management agencies, and law enforcement agencies and enforcers to provide responses prompt and appropriate to emergency situations, or regulatory violations are imposed.

13.11.3 Recommendations for industry stakeholders

- Integrate Privacy-by-Design and Compliance-by-Design:
  The manufacturers of the technology and the individuals controlling the UAVs ought to build in the data security and regulatory measures

in the blockchain systems since their inception, as these will ensure the data safety and privacy regulations in the aviation industry.

- Early Alignment with Regulatory Standards:
  The industry is also supposed to collaborate with the regulators to make sure that smart contracts, flight logs, and operating data meet legal and safety criteria and minimize legal liability and gain trust.
- Operational Transparency and ESG Integration:
  The support of ESG reporting and SDGs compliance can be offered with the assistance of blockchain that makes it possible to monitor UAV activities, maintenance records, and sustainability rates.

13.12 CONCLUSION: OPERATIONALIZING LEGAL PRINCIPLES THROUGH BLOCKCHAIN

The process of changing the principles of governance according to which the abstract principles of regulation transform into the evidence-based and enforceable forms of governance is known as the introduction of the blockchain technology into the UAV governance. The perception of such a concept as blockchain as an enabler technology should never take place, but rather as a regulatory infrastructure, which can operationalize legal time or legal ideas of transparency, accountability, and proportionality [50]. In a more or less international context of locating the developing UAV and data protection facilities in India, this chapter explains how India may contribute in the establishment of governance norms that may be applied at the emerging economies. The two-tier system of governance makes certain that the domestic regulatory autonomy is realized but the international harmonization, law certainty, and level of operation is facilitated.

Moreover, blockchain governance facilitates sustainable UAV governance by enhancing the carbon footprint governance, effectiveness of disaster management and environmental sustainability, and responsibility thus fostering the SDGs. Finally, the chapter is a practitioner-led, multidimensional roadmap of the manner blockchain must be incorporated in the UAV governance and innovation, legal accountability, and sustainability endeavors in a future-oriented, sustainable UAV ecosystem.

REFERENCES

[1] Kshetri, N. (2017). Can blockchain strengthen the internet of things? *IT Professional*, 19(4), 68–72.
[2] Christidis, K., & Devetsikiotis, M. (2016). Blockchains and smart contracts for the internet of things. *IEEE Access*, 4, 2292–2303.
[3] Fernández-Caramés, T. M., & Fraga-Lamas, P. (2018). A review on the use of blockchain for the Internet of Things. *IEEE Access*, 6, 32979–33001.
[4] Yaga, D., Mell, P., Roby, N., & Scarfone, K. (2018). Blockchain technology overview. NIST Internal Report 8202.

[5] Ølnes, S., Ubacht, J., & Janssen, M. (2017). Blockchain in government: Benefits and implications of distributed ledger technology for information sharing. *Government Information Quarterly*, 34(3), 355–364. https://doi.org/10.1016/j.giq.2017.09.007

[6] Casino, F., Dasaklis, T. K., & Patsakis, C. (2019). A systematic literature review of blockchain-based applications: Current status, classification and open issues. *Telematics and Informatics*, 36, 55–81. https://doi.org/10.1016/j.tele.2018.11.006

[7] Saberi, S., Kouhizadeh, M., Sarkis, J., & Shen, L. (2019). Blockchain technology and its relationships to sustainable supply chain management. *International Journal of Production Research*, 57(7), 2117–2135. https://doi.org/10.1080/00207543.2018.1533261

[8] Dorri, A., Kanhere, S. S., Jurdak, R., & Gauravaram, P. (2017). Blockchain for IoT security and privacy: The case study of a smart home. In *2017 IEEE International Conference on Pervasive Computing and Communications Workshops (PerCom Workshops)*, 618–623. https://doi.org/10.1109/PERCOMW.2017.7917634

[9] Tian, F. (2016). An agri-food supply chain traceability system for China based on RFID & blockchain technology. In *2016 13th International Conference on Service Systems and Service Management (ICSSSM)*, 1–6. https://doi.org/10.1109/ICSSSM.2016.7538424

[10] Reyna, A., Martín, C., Chen, J., Soler, E., & Díaz, M. (2018). On blockchain and its integration with IoT. Challenges and opportunities. *Future Generation Computer Systems*, 88, 173–190. https://doi.org/10.1016/j.future.2018.05.046

[11] Aggarwal, S., Chaudhary, R., Aujla, G. S., Kumar, N., Choo, K. K. R., & Zomaya, A. Y. (2019). Blockchain for smart communities: Applications, challenges and opportunities. *Journal of Network and Computer Applications*, 144, 13–48. https://doi.org/10.1016/j.jnca.2019.06.018

[12] Biswas, K., & Muthukkumarasamy, V. (2016). Securing smart cities using blockchain technology. In *2016 IEEE 18th International Conference on High Performance Computing and Communications*, 1392–1393. https://doi.org/10.1109/HPCC-SmartCity-DSS.2016.0198

[13] Singh, S., Sharma, P. K., Yoon, B., Shojafar, M., Cho, G. H., & Ra, I. H. (2020). Convergence of blockchain and artificial intelligence in IoT network for the sustainable smart city. *Sustainable Cities and Society*, 63, 102364. https://doi.org/10.1016/j.scs.2020.102364

[14] Salah, K., Rehman, M. H. U., Nizamuddin, N., & Al-Fuqaha, A. (2019). Blockchain for AI: Review and open research challenges. *IEEE Access*, 7, 10127–10149. https://doi.org/10.1109/ACCESS.2018.2890507

[15] Zheng, Z., Xie, S., Dai, H. N., Chen, X., & Wang, H. (2018). Blockchain challenges and opportunities: A survey. *International Journal of Web and Grid Services*, 14(4), 352–375. https://doi.org/10.1504/IJWGS.2018.095647

[16] Khan, M. A., & Salah, K. (2018). IoT security: Review, blockchain solutions, and open challenges. *Future Generation Computer Systems*, 82, 395–411. https://doi.org/10.1016/j.future.2017.11.022

[17] Fernández-Caramés, T. M., Froiz-Míguez, I., Blanco-Novoa, O., & Fraga-Lamas, P. (2019). Enabling the Internet of mobile crowdsourcing health things: A mobile fog computing, blockchain and IoT based continuous glucose

monitoring system for diabetes mellitus research and care. *Sensors*, 19(15), 3319. https://doi.org/10.3390/s19153319
[18] Zhao, G., Liu, S., Lopez, C., Lu, H., Elgueta, S., Chen, H., & Boshkoska, B. M. (2019). Blockchain technology in agri-food value chain management: A synthesis of applications, challenges and future research directions. *Computers in Industry*, 109, 83–99. https://doi.org/10.1016/j.compind.2019.04.002
[19] Lohmer, J., & Lasch, R. (2020). Blockchain in operations management and manufacturing: Content analysis and research agenda. *Computers & Industrial Engineering*, 143, 106400. https://doi.org/10.1016/j.cie.2020.106789
[20] Kouhizadeh, M., Saberi, S., & Sarkis, J. (2021). Blockchain technology and the sustainable supply chain: Theoretically exploring adoption barriers. *International Journal of Production Economics*, 231, 107831. https://doi.org/10.1016/j.ijpe.2020.107831
[21] Hassija, V., Chamola, V., Saxena, V., Jain, D., Goyal, P., & Sikdar, B. (2019). A survey on IoT security: Application areas, security threats, and solution architectures. *IEEE Access*, 7, 82721–82743. https://doi.org/10.1109/ACCESS.2019.2924045
[22] Mylrea, M., & Gourisetti, S. N. G. (2017). Blockchain for smart grid resilience: Exchanging distributed energy at speed, scale and security. In *2017 Resilience Week (RWS)*, Wilmington, DE, 18–23. https://doi.org/10.1109/RWEEK.2017.8088642
[23] Gai, K., Wu, Y., Zhu, L., Qiu, M., & Shen, M. (2019). Privacy-preserving energy trading using consortium blockchain in smart grid. *IEEE Transactions on Industrial Informatics*, 15(6), 3548–3558. https://doi.org/10.1109/TII.2019.2893433
[24] Jamil, F., Hang, L., Kim, K., & Kim, D. (2019). A novel medical blockchain model for drug supply chain integrity management in a smart hospital. *Electronics*, 8(5), 505. https://doi.org/10.3390/electronics8050505
[25] Angelis, J., & Ribeiro da Silva, E. (2019). Blockchain adoption: A value driver perspective. *Business Horizons*, 62(3), 307–314. https://doi.org/10.1016/j.bushor.2018.12.001
[26] Bodkhe, U., Tanwar, S., Parekh, K., Khanpara, P., Tyagi, S., Kumar, N., & Alazab, M. (2020). Blockchain for industry 4.0: A comprehensive review. *IEEE Access*, 8, 79764–79800. https://doi.org/10.1109/ACCESS.2020.2988579
[27] Rejeb, A., Keogh, J. G., & Treiblmaier, H. (2019). Leveraging the internet of things and blockchain technology in supply chain management. *Future Internet*, 11(7), 161. https://doi.org/10.3390/fi11070161
[28] Perboli, G., Musso, S., & Rosano, M. (2018). Blockchain in logistics and supply chain: A lean approach for designing real-world use cases. *IEEE Access*, 6, 62018–62028. https://doi.org/10.1109/ACCESS.2018.2875782
[29] Cole, R., Stevenson, M., & Aitken, J. (2019). Blockchain technology: Implications for operations and supply chain management. *Supply Chain Management: An International Journal*, 24(4), 469–483. https://doi.org/10.1108/SCM-09-2018-0309
[30] Ghosh, A., Gupta, S., Dua, A., & Kumar, N. (2020). Security of cryptocurrencies in blockchain technology: State-of-art, challenges and future prospects. *Journal of Network and Computer Applications*, 163, 102635. https://doi.org/10.1016/j.jnca.2020.102635

[31] Benchoufi, M., & Ravaud, P. (2017). Blockchain technology for improving clinical research quality. *Trials*, 18(1), 335. https://doi.org/10.1186/s13063-017-2035-z
[32] Hölbl, M., Kompara, M., Kamišalić, A., & Nemec Zlatolas, L. (2018). A systematic review of the use of blockchain in healthcare. *Symmetry*, 10(10), 470. https://doi.org/10.3390/sym10100470
[33] Aste, T., Tasca, P., & Di Matteo, T. (2017). Blockchain technologies: The foreseeable impact on society and industry. *Computer*, 50(9), 18–28. https://doi.org/10.1109/MC.2017.3571064
[34] Beck, R., Müller-Bloch, C., & King, J. L. (2018). Governance in the blockchain economy: A framework and research agenda. *Journal of the Association for Information Systems*, 19(10), 1020–1034. https://doi.org/10.17705/1jais.00518
[35] Buterin, V. (2014). A next-generation smart contract and decentralized application platform. *Ethereum White Paper*, 3(37).
[36] Tapscott, D., & Tapscott, A. (2017). How blockchain will change organizations. *MIT Sloan Management Review*, 58(2), 10–13. https://doi.org/10.7551/mitpress/11645.003.0010
[37] Crosby, M., Pattanayak, P., Verma, S., & Kalyanaraman, V. (2016). Blockchain technology: Beyond bitcoin. *Applied Innovation*, 2(6–10), 71. https://doi.org/10.1515/9783110488951-003
[38] Pilkington, M. (2016). Blockchain technology: Principles and applications. In *Research Handbook on Digital Transformations*, 225–253. https://doi.org/10.4337/9781784717766.00019
[39] Morabito, V. (2017). *Business Innovation Through Blockchain: The B3 Perspective*. Springer International Publishing.
[40] Nakamoto, S. (2008). Bitcoin: A peer-to-peer electronic cash system. *Decentralized Business Review*, 21260. https://doi.org/10.2139/ssrn.3440802
[41] Haque, A. B., Bhushan, B., & Dhiman, G. (2022). Conceptualizing smart city applications: Requirements, architecture, security issues, and emerging trends. *Expert Systems*, 39(5), e12753. https://doi.org/10.1111/exsy.12753
[42] Gupta, R., Tanwar, S., Tyagi, S., & Kumar, N. (2020). Machine learning models for secure data analytics: A taxonomy and threat model. *Computer Communications*, 153, 406–440. https://doi.org/10.1016/j.comcom.2020.02.008
[43] Javed, M. U., Rehman, M., Javaid, N., Aldegheishem, A., Alrajeh, N., & Tahir, M. (2020). Blockchain-based secure data storage for distributed vehicular networks. *Applied Sciences*, 10(6), 2011. https://doi.org/10.3390/app10062011
[44] Lacity, M. C. (2018). Addressing key challenges to making enterprise blockchain applications a reality. *MIS Quarterly Executive*, 17(3), 201–222.
[45] Karame, G., & Androulaki, E. (2016). *Bitcoin and Blockchain Security*. Artech House.
[46] Iansiti, M., & Lakhani, K. R. (2017). The truth about blockchain. *Harvard Business Review*, 95(1), 118–127.
[47] Chamola, V., Hassija, V., Gupta, V., & Guizani, M. (2020). A comprehensive review of the COVID-19 pandemic and the role of IoT, drones, AI, blockchain, and 5G in managing its impact. *IEEE Access*, 8, 90225–90265. https://doi.org/10.1109/ACCESS.2020.2992341

[48] Dinh, T. N., & Thai, M. T. (2018). AI and blockchain: A disruptive integration. *Computer*, 51(9), 48–53. https://doi.org/10.1109/MC.2018.3620971
[49] Novo, O. (2018). Blockchain meets IoT: An architecture for scalable access management in IoT. *IEEE Internet of Things Journal*, 5(2), 1184–1195. https://doi.org/10.1109/JIOT.2018.2812239
[50] Anand, A., McKibbin, M., & Pichel, F. (2020). Colored coins: Bitcoin, blockchain, and land administration. In *Proceedings of the Annual World Bank Conference on Land and Poverty*, 1–18. https://doi.org/10.1596/33726

Index

Pages in *italics* refer to figures and pages in **bold** refer to tables.

For Product Safety Concerns and Information please contact our EU representative GPSR@taylorandfrancis.com
Taylor & Francis Verlag GmbH, Kaufingerstraße 24, 80331 München, Germany

www.ingramcontent.com/pod-product-compliance
Ingram Content Group UK Ltd.
Pitfield, Milton Keynes, MK11 3LW, UK
UKHW021703300926
475953UK00017B/45
9781041222880